MINERALS AND MINERALOIDS IN MARINE SEDIMENTS

AN OPTICAL IDENTIFICATION GUIDE

MINERALS AND MINERALOIDS IN MARINE SEDIMENTS

An Optical Identification Guide

R. G. ROTHWELL

Institute of Oceanographic Sciences,
Wormley, Godalming, UK

ELSEVIER APPLIED SCIENCE
LONDON and NEW YORK

ELSEVIER SCIENCE PUBLISHERS LTD
Crown House, Linton Road, Barking, Essex IG11 8JU, England

Sole Distributor in the USA and Canada
ELSEVIER SCIENCE PUBLISHING CO., INC.
655 Avenue of the Americas, New York, NY 10010, USA

WITH 8 TABLES AND 150 ILLUSTRATIONS

Softcover reprint of the hardcover 1st edition 1989

British Library Cataloguing in Publication Data

Rothwell, R. G.
Minerals and mineraloids in marine sediments.
1. Marine sediments. Minerals
I. Title
551.46'083

ISBN-13: 978-94-010-7002-7 e-ISBN-13: 978-94-009-1133-8
DOI: 10.1007/978-94-009-1133-8

Library of Congress CIP Data

Rothwell, R. G.
Minerals and mineraloids in marine sediments.

Bibliography: p.
Includes index.
1. Marine sediments. 2. Mineralogy. I. Title.
GC380.15.R68 1989 549.9162 89-11612

FOREWORD

Over 60% of the Earth's surface is covered with deep marine sediments, however, until the early 1980s, no comprehensive text books appeared to support the rapid expansion in the study of these sediments. While the whole field of marine geology has expanded enormously and entirely new disciplines, such as paleoceanography, have been developed, there remains a lack of reference texts on study techniques that investigators in the marine community can turn to. *Minerals and Mineraloids in Marine Sediments* is an optical identification guide that I believe will become a standard reference text for use in the microscope analysis of marine sediments and sedimentary rocks.

The systematic collection of sediment cores from the deep ocean floor began in earnest with the Swedish Deep Sea Expedition, 1947–1948. Much of the microscopic examination of the sediments collected in these piston cores (10 m+ long) was conducted on separated grain mounts or thin sections of impregnated sediments. By the late 1960s a simpler technique of examining a mounted smear of the cored silt and clay size sediment on a microscope slide had become standard practice in American oceanographic institutions. This semi-quantitative technique became the standard tool used in core description aboard *Glomar Challenger* through the 15 years of the Deep Sea Drilling Project (DSDP), 1968–1983. Visual percentage estimates of biogenic and mineral components were made using petrologic microscopes. Steadily improving deep sea sediment classification schemes were applied to provide names for sediment and rock sequences that were drilled at 624 DSDP site locations worldwide. Hundreds of sedimentologists used this method through the 96 cruises of DSDP and many continued its use on marine cores in other projects. Always recognised were the built-in limitations of the method introduced through operator error or bias in visual estimation. Even experienced sedimentary petrologists require a period of self-teaching, examining slides of representative sediment types, to overcome the differences between smear slide analysis and thin section or grain mount examination.

The success of the simple, rapid and cheap smear slide technique comes about because:

(1) it is a means of quickly identifying sediment types, such that only rarely does later sample analysis using quantitative X-ray mineralogical and other techniques require a sediment name to be changed.
(2) it can be used to recognise relative changes in mineral or biogenic components through a sequence.

The method was continued when the successor to DSDP, the international Ocean Drilling Program (ODP) was launched in 1985. Basic core description aboard *JOIDES Resolution* continues to be associated with the examination of smear slides.

In my view the technique is much underused outside the Ocean Drilling Program. Many have used it, for example, in the rapid initial examination of marine sedimentary rocks collected in seafloor dredge hauls and even from outcrops on land. As in scientific drilling (ODP and DSDP), smear slides can be used to examine downhole variability in industry boreholes through from unconsolidated silts and clays to semi-lithified rocks. Soils may even be examined in this way, although terrestrial microscopic components would differ greatly from those described in this book.

During DSDP and the initial years of ODP there was never a satisfactory manual or reference text to aid in microscopic identification of the smear slide components, despite its importance in basic data collection. This fact is likely to have contributed to inherent problems of operator bias and error. Many sedimentologists in the marine community collected together reference sets of slides and notes and some assembled their own personal reference manuals but, until this text, there has been no published handbook. We were all forced to use reference texts that were designed for thin section mineral identification such as Kerr's *Optical Mineralology* (1959) or Deer, Howie and Zussman's *Introduction to the Rock Forming Minerals* (1966). A number of false starts were made on manuals by workers at DSDP (Scripps Institution of Oceanography) and ODP (Texas A & M University) and also at some other oceanographic institutions.

In the early 1970s, as a post doctoral visiting scientist at DSDP, La Jolla, I collaborated with Oscar Weser on an initial manual for the Project and also gathered together a reference set of slides for the *Glomar Challenger*. The effort was never fully completed—everyone was too involved in the exciting discoveries that came almost monthly from the Project at that time (!). Subsequently at the Institute of

Oceanographic Sciences (IOSDL) Wormley, I introduced a large number of colleagues and postgraduate students to the method, always intending to bring together the reference material into a complete manual. Collaboration with Guy Rothwell on a number of sediment core-based programmes eventually brought the project nearer to fruition.

Guy diligently amassed an immense reference set of literature and sample material on sedimentary components. He is an extremely gifted sedimentary petrologist and has built up an impressive knowledge of deep sea sediments. He first completed a laboratory smear slide manual for the Institute of Oceanographic Sciences. He was very helpful later on when I was seeking reference literature for the core laboratory on *JOIDES Resolution* during my period as head of the Science Operations Group at the Ocean Drilling Program, College Station, Texas. He was encouraged through this experience, and by his colleagues at IOSDL, to publish a reference handbook. The result is *Minerals and Mineraloids in Marine Sediments—an optical identification guide.*

Smear slide analysis is simple, rapid and cheap. It requires merely access to a good quality petrologic microscope and a laboratory hot plate supported by the cheapest of raw materials (microscope slides, cover slips, mounting medium, labels, a distilled water bottle and toothpicks). The sample required is so small (head-of-a-toothpick size) that even archive material can be used. Now that this guide is published there is no longer the necessity for the operator to be entirely 'self taught'.

The appeal of Guy Rothwell's book will primarily be as a handy reference and comprehensive text for those of us in the sedimentological community who continue to work on marine cores, ODP and other borehole samples and materials from marine sedimentary rock outcrops. It incorporates some magnificent microphotography and has an impressive bibliography. This guide is long overdue!

Robert B. Kidd
Professor of Geology,
University College of Wales,
Cardiff, UK

PREFACE

The use of smear slides to analyse core lithologies often forms the main investigative technique in the practical description of marine sediments. However, to the geologist used to the standard thickness of conventional thin sections, even common minerals may sometimes appear unfamiliar when seen in smear slides. Firstly, in smear slides, mineral grains will not be of uniform thickness, thus criteria such as birefringence and optical character are more difficult to use for identification purposes. Secondly, the mineral grains will be unconstrained, being dispersed through a clear mounting medium. Consequently minerals in which particular forms dominate their habit, such as micas, will tend to settle in preferred orientations on the slide, affecting the criteria useful for recognition. Further, a range of authigenic minerals can occur in marine sediments. These can often reveal much concerning sedimentary history and may be quite unfamiliar to the geologist trained on terrestrial sediments.

To the author's knowledge, no manual presently exists which serves as an identification guide to minerals commonly found in marine sediments. This book tries to fulfil that need but it is meant to be more than just a simple identification guide. An extensive literature was reviewed during its preparation and a number of mineralogical studies are referred to in the text which illustrate the application and usefulness of mineralogical analyses of marine sediments in a wide range of research.

The main text of this book deals only with the detrital, biogenic, volcanogenic and authigenic minerals which may be encountered in the analysis of smear slides. There is little discussion of microfossils, partly due to limitations on space and to the fact that there are several good micropalaeontological texts available. However, an illustrated key to the main microfossil groups, with recommendations for further reading, is included as an appendix. This will familiarise readers with the main types of microfossils likely to be seen in marine sediments.

ACKNOWLEDGEMENTS

This publication has only been possible through the work of a large number of sedimentologists. I should like to thank Professor R. B. Kidd for writing the Foreword. He originally conceived the need for a work of this kind and also reviewed the text. I should also like to thank Ms Anne Graham and the Curators at the Ocean Drilling Program for their help in supplying sediment samples. I also thank Mr Michael Conquer for his careful printing of the photomicrographs. He spent many hours in the darkroom and never failed to produce excellent results. The text also benefitted from criticisms and suggestions made by Dr Phil Hill to whom I also express my thanks. I thank Dr Phil Weaver for reviewing Appendix 5.

Special acknowledgement is made to the following colleagues, authors, publishers and organisations who provided additional illustrative material: Drs Michel Hoffert, Thomas Donnelly, Gino Nalli, Peter Borella, C. Adelseck, Annika Sanfilippo, W. R. Riedel, Billy Glass, F. T. Kyte, Wolfgang Berger, Ulrich Von Rad, Mary Ann Holmes, Jim Breza, Sherwood W. Wise, Walter E. Dean, Margaret Leinen, Dorrik Stow, Jim Mazzullo, Audrey Meyer, Robert B. Kidd, Quentin Huggett, Phillip Weaver, Sally Horn, Pat Doyle, Jean–Pierre Caulet, Peta Mudie, Macmillan Journals, the Society of Economic Paleontologists and Mineralogists, the Ocean Drilling Program and the Deep Sea Drilling Project.

In addition I wish to thank those who typed the many drafts of the manuscript, especially Mrs G. Mabley, Mrs J. Jones, Mrs A. Williams, Mrs M. Campbell and Mrs J. Breach.

CONTENTS

PART I

THE MINERALOGY OF MARINE SEDIMENTS

THE SMEAR SLIDE METHOD

THE MINERALOGY OF MARINE SEDIMENTS

THE COMPOSITION OF MARINE SEDIMENTS

The individual sedimentary grains that make up marine sediments may be classified according to their origin.

Detrital or Terrigenous

Those grains that were originally formed on the continents and have subsequently been transported into the ocean. River transport is the most important mechanism in which terrigenous minerals are passed to the ocean realm. Milliman & Meade (1983) estimate that about $12–13 \times 10^9$ tonnes of suspended sediment are supplied to the oceans by rivers annually and that a further $1–2 \times 10^9$ tonnes are supplied by bedload and flood discharges. If this amount were evenly distributed on the ocean floor, it alone would account for a sedimentation rate of over 3 cm per 1000 year (Dietrich *et al.*, 1980). However, the actual increase in deep-sea sediments is much lower, as much of the terrigenous material is deposited on the continental shelf and rise. Other important pathways by which detrital minerals are passed to the deep sea are ice-rafting and aeolian transport, although these mechanisms are much less important volumetrically than river transport. Garrels & MacKenzie (1971) give an estimate of $0{\cdot}06 \times 10^9$ tonnes/year for the total mass of material conveyed by aeolian transport to the oceans and 2×10^9 tonnes/year for ice-rafting. The most common detrital mineral in deep-ocean sediments is quartz derived from the weathering of continental rocks.

Biogenic

Those minerals formed through biological activity in the sea. Such grains consist largely of residua of microfossil tests. Calcite, which is a common test material produced by a number of microfossil groups (e.g.

foraminifera and coccolithophores), is the most common biogenic component seen in deep-sea sediments. However, in areas of high productivity biogenic silica (e.g. radiolaria and diatoms) may be important and may dominate the biogenic component in deep basins where carbonate solution is high.

Volcanogenic

Those minerals or mineraloids introduced directly onto the seafloor through submarine volcanism or derived from terrestrial volcanism. Subaerial volcanic eruptions may provide significant input to marine sedimentation via ash-falls, pumice-rafting and pyroclastic flows. Altered volcanic ash is a major component of some marine sediments and may be a source of manganese and iron precipitates. Examples of common volcanogenic grains in marine sediments are volcanic glass and some feldspars and pyroxenes.

Cosmogenic

Those particles having an extraterrestrial origin (e.g. iron–nickel spherules and microtektites). Extraterrestrial material is most abundant in areas of slowly accumulating sediments far from the continents, such as the Central Pacific. Extraterrestrial sources are estimated to supply only $0{\cdot}0036 \times 10^9$ tonnes of material to the oceans annually (Garrels & MacKenzie, 1971).

Authigenic

Those minerals formed directly in the ocean or in ocean sediments by means of reactions involving seawater or interstitial solutions. Equivalent terms are hydrogenous (Goldberg, 1954) and halmeic (Arrhenius, 1963). Halmyrolysis is a term used to indicate such reactions when they occur in the water column or during the preburial stage of diagenesis. Authigenic minerals may be in part biochemically precipitated. Such minerals include manganese, iron and other metal oxides, barium sulphate, aluminosilicates and, in shallow water, phosphates and glauconite. The most important of these are the ferromanganese nodules.

Hydrothermal

Those minerals formed through precipitation from hydrothermal solutions injected into seawater through the seafloor (Bonatti, 1981). The most important of these are the metalliferous sediments that have been described from oceanic spreading centres, fracture zones, ophiolite complexes and marginal basins. Mineralogically, these deposits fall into three groups:

(a) Metal oxides/hydroxides—mixtures of hydrated iron and manganese oxides and hydroxides (e.g. goethite, birnessite and todorokite).
(b) Metal silicates—principally iron-rich smectites.
(c) Metal sulphides—such as sphalerite, pyrite and chalcopyrite.

Besides these a variety of other minerals have been ascribed to a hydrothermal origin in certain instances (e.g. aragonite, calcite and zeolites).

Hydrothermal minerals may occur as 'concentrated deposits' in which they form the dominant component (>50%) (Fig. 1), these generally

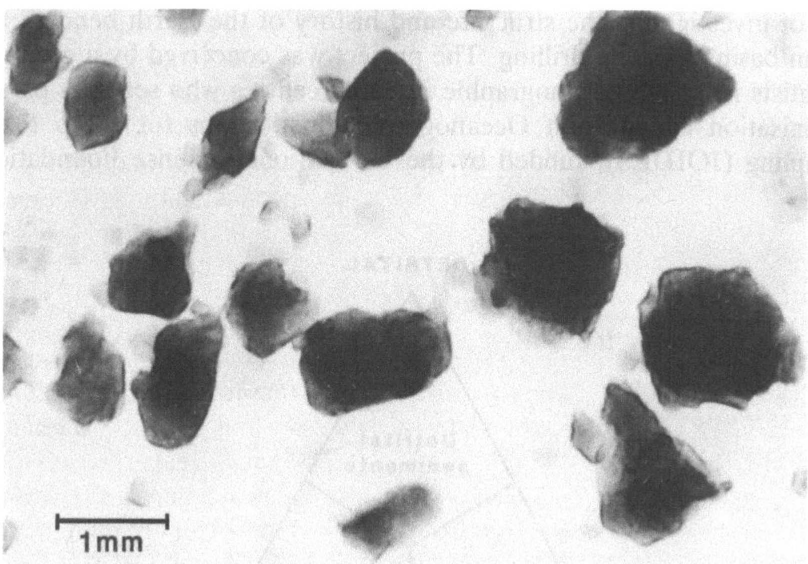

Fig. 1. An example of a concentrated hydrothermal deposit: hydrothermal granules, green to brownish-black in plane-polarised light, from the Pleistocene of the Galapagos Rift, Panama Basin (00°35.33′N, 86°07.81′W). X-ray diffraction studies have shown that the granules have a clay structure characteristic of the smectite group (Hoffert *et al.*, 1980). DSDP Leg 54, Site 424A, core 2, section 3, 142–144 cm. (Courtesy of Dr Michel Hoffert and the Deep Sea Drilling Project.)

occur close to sites of hydrothermal activity or as 'diluted deposits' where the hydrothermal minerals are diluted by terrigenous, biogenic and/or authigenic material.

SEDIMENT CLASSIFICATION

Volcanogenic, cosmogenic, authigenic and hydrothermal material, although they may be of local importance, are usually minor constituents of most deep-sea sediments. In most marine sediments either the terrigenous or biogenic component (calcareous or siliceous) is dominant. Thus most marine sediments can be described within a three-component system, the end-members being terrigenous, biogenic–calcareous and biogenic–siliceous (Fig. 2). Most sediment classifications use this three-component system, including those devised by Deep Sea Drilling Project/Ocean Drilling Program scientists.

The Deep Sea Drilling Project (DSDP) was initiated in 1968 with the aim of investigating the structure and history of the Earth beneath the ocean basins by deep drilling. The project was conceived by a group of scientists from US oceanographic research centres who set up a parent organisation called Joint Oceanographic Institutions for Deep Earth Sampling (JOIDES), funded by the US National Science Foundation.

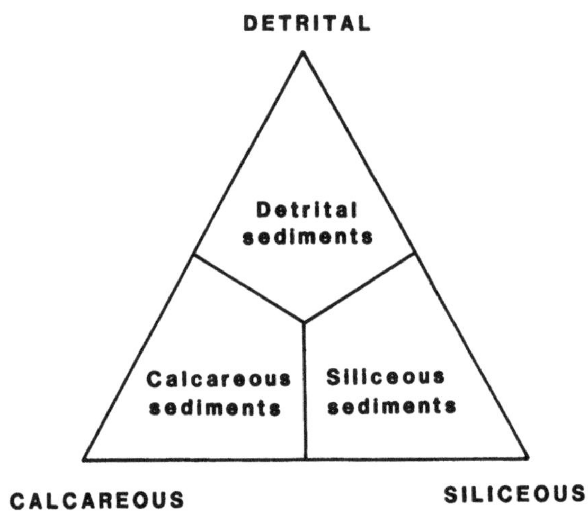

Fig. 2. General classification of marine sediments within a three-component system, the end-members being detrital, biogenic–calcareous and biogenic–siliceous sediments.

Using the dynamically-positioned drillship 'Glomar Challenger', the DSDP had drilled 1092 holes at 624 sites worldwide and recovered 96 km of core by November 1983 (see Francis, 1986, for a review of DSDP/ODP drilling technology). However, by 1975, the success of the project was such that an international phase was initiated. Research organisations from Britain, France, West Germany, the USSR and Japan became members of JOIDES and participated in the Deep Sea Drilling Project. This International Phase of Ocean Drilling (IPOD) was superseded by a new international research program—the Ocean Drilling Program (ODP), commencing operations at sea in 1985 with a new drillship 'Joides Resolution' which had improved capabilities. The successes and scientific benefits of the ODP and its precursors are far too numerous to discuss here but the program has led to great advances in our understanding of the processes of plate tectonics, of the Earth's crustal structure and composition, of past changes in climate and of conditions in ancient oceans. Deep drilling has and will lead to a greater comprehension of the evolution of the Earth.

The JOIDES Sedimentary Petrology and Physical Properties Panel (SP4) has devised a three-component sediment classification which has been the most widely used by DSDP/ODP scientists. This is published in the introduction of the DSDP Initial Reports volume for Leg 42, Part 2 (Ross *et al.*, 1978). Recently Dean *et al.* (1985) have produced a more detailed classification within the confines of the three main components, especially suited to fine-grained sediments (Fig. 3). The latest ODP classification scheme (Mazzullo *et al.*, 1987) incorporates many of the suggestions proposed by Dean *et al.* (1985) but also includes a classification of neritic sediments. This classification was approved by the JOIDES Sediments and Ocean History Panel in 1987 for shipboard use and is intended to supersede earlier schemes. Their scheme is descriptive rather than genetic, classifying sediments on their compositions and textures rather than their assumed or inferred genesis.

Mazzullo *et al.* (1987) define two basic sediment types:

(a) Granular sediment—composed of discrete organic grains (e.g. foraminifer and radiolarian tests) or inorganic grains (e.g. quartz, rock fragments, tephra), deposited by physical or organic processes.
(b) Chemical sediment—composed of minerals formed by inorganic processes such as compaction and heating of carbonaceous plant debris, precipitation from solution or colloidal suspension, deposition of insoluble precipitates or recrystallisation of detrital

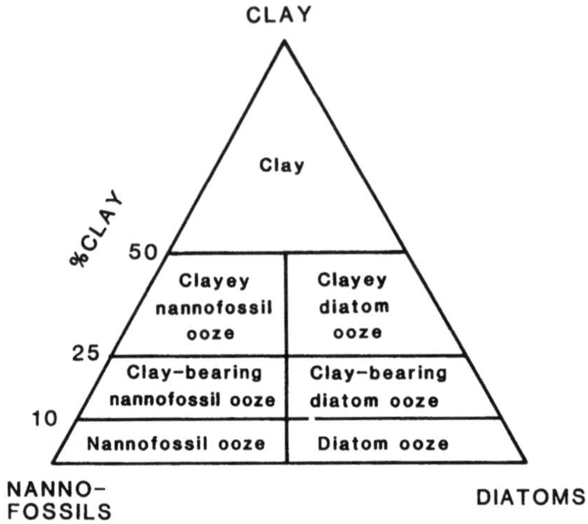

Fig. 3. Classification of fine-grained sediments within a three-component system, the end-members being clay, nannofossils and diatoms (from Dean *et al.,* 1985). (Courtesy of Dr Walter Dean and the Society of Economic Paleontologists and Mineralogists).

evaporites and calcareous or siliceous biogenic debris. Chemical sediments typically have crystalline textures.

Examples of granular sediments are foraminiferal ooze, nannofossil chalk, oolitic grainstone and quartz sand. Some examples of chemical sediments are coal, halite and gypsum. Mazzullo *et al.* (1987) have devised separate sediment schemes for granular and chemical sediments (Tables 1, 2 and 3). Their classification of granular sediments defines five major classes of granular sediments: pelagic, neritic, siliciclastic, volcaniclastic and mixed (Fig. 4), defined on the dominant grain type present (Table 4).

The composition, texture and proportions of the different sedimentary grains making up a sediment are estimated through smear slide analysis (or thin section analysis, if the sediment is consolidated) using a petrological microscope; although, in certain instances, the composition of some samples may be determined by more accurate methods such as X-ray diffraction.

Sediment samples are given a main name which defines the sediment type; a major modifier(s) describing the composition and textures of the dominant components (i.e. components present in abundances of 25–100%) plus a minor modifier(s) describing the composition and textures of components present in abundances of between 10 and 25%.

TABLE 1 Classification of Granular Sediment (Mazzullo *et al.*, 1987)

1. PELAGIC SEDIMENTS

Defined as being composed of >60% pelagic and neritic grains and <40% siliciclastic and volcanic grains. Pelagic grains are dominant.

Principal names (describes composition and degree of consolidation).

Ooze: unconsolidated calcareous and/or siliceous pelagic sediment.
Chalk: firm, semi-consolidated pelagic sediment composed mainly of calcareous pelagic grains.
Limestone: hard, indurated pelagic sediment composed predominantly of calcareous pelagic grains.
Radiolarite, diatomite and spiculite: firm siliceous pelagic sediment composed predominantly of radiolarians or diatoms or sponge spicules.
Chert: hard, indurated sediment composed mainly of siliceous pelagic grains.

Major modifiers always precede the principal name and describe the composition and textures of grains present in abundances of 25–100%. In addition major modifiers can be used to describe grain fabric (e.g. 'grain-supported', 'matrix-supported', 'imbricated'), grain shape (e.g. 'rounded', 'subrounded', 'subangular', 'angular') and sediment colour.

Common major modifiers are 'nannofossil', 'foraminifer(-al)', 'calcareous', 'diatom(-aceous)', 'radiolarian', 'spicule(-ar)', and 'siliceous'.

Minor modifiers always follow the principal name (preceded by 'with') and describe the composition and textures of grains present in abundances of 10–25%.

2. NERITIC SEDIMENTS

Defined as being composed of >60% pelagic and neritic grains and <40% siliciclastic and volcaniclastic grains but neritic grains are dominant.

Principal names (describes texture and fabric) (terms from Dunham, 1962).

Boundstone: components organically bound during deposition.
Grainstone; grain-supported fabric, no mud, grains <2 mm in size.
Packstone: grain-supported fabric with intergranular mud, grains <2 mm in size.

(continued)

TABLE 1—*contd.*

Wackestone: mud-supported fabric but containing >10% grains, grains <2 mm in size.
Mudstone: mud-supported fabric but containing <10% grains.
Floatstone: matrix-supported fabric, grains >2 mm in size.
Rudstone: grain-supported fabric, grains >2 mm in size.

Major and minor modifiers
Common major and minor modifiers used in describing neritic sediments are:

Ooid (or oolitic): spherical or elliptical non-skeletal particles, <2 mm in diameter, having a central nucleus surrounded by an accretionary shell showing a concentric or radial structure.
Bioclast (-ic): fragmental skeletal material, specific names such as 'molluscan' or 'algal' can also be used.
Pellet (-al): faecal grains excreted by molluscs, echinoderms and other groups of invertebrate animals.
Intraclast (-ic): reworked carbonate rock fragments.
Pisolite (pisolitic): large ooliths (>2 mm in diameter).
Peloid (-al): micritised carbonate particle of uncertain origin.

Other modifiers used to describe the composition of neritic sediments include 'calcareous', 'dolomitic', 'aragonitic' and 'sideritic'.

3. *SILICICLASTIC SEDIMENT*
Defined as comprising >60% siliciclastic and volcaniclastic grains and <40% pelagic and neritic grains. Siliciclastic grains are present in a greater proportion than volcaniclastic grains.

Principal names (describes grain size, according to Wentworth scale (Wentworth, 1922; Table 2).

Gravel: individual grains >2 mm in size.
Conglomerate: lithified gravel having rounded clasts.
Breccia: lithified gravel having angular clasts.
Sand (-stone, when lithified): individual grains 0·06–2 mm in size. Sand may be described as very fine, fine, medium, coarse or very coarse according to its size—see Table 2.
Silt (-stone, when lithified): individual grains 0·004–0·06 mm in size. May be described as very fine, fine, medium or coarse according to size (Table 2).
Clay (-stone when lithified): individual grains less than 0·004 mm in size.
Shale: fissile siltstone or claystone.

When a sediment is composed of two or more grain size groups, they are listed as principal names in order of increasing abundance.

TABLE 1—*contd.*

Major and minor modifiers
The mineralogy of siliciclastic sediments is described by using modifiers such as 'quartz', 'feldspar', 'glauconite', 'lithic', 'sapropelic', 'zeolitic', etc. The source of rock fragments, particularly in gravels, conglomerates and breccias, may also sometimes be described by using modifiers such as 'volcanic', 'sed-lithic', 'meta-lithic', 'gneissic', 'basaltic', etc.

4. *VOLCANICLASTIC SEDIMENTS*
Defined as sediments composed of >60% siliciclastic and volcaniclastic grains and <40% pelagic and neritic grains: volcaniclastic grains being dominant.

Principal names (describes texture).

Volcanic breccia: component volcaniclastic grains are >64 mm in diameter.
Volcanic lapilli: component volcaniclastic grains are 2–64 mm in diameter.
Volcanic ash (tuff, when lithified): component volcaniclastic grains are <2 mm in diameter.

Major and minor modifiers
The composition of the volcaniclastic grains is described by modifiers such as 'vitric' (glass), 'lithic' (rock fragments) and 'crystal' (mineral crystals). The composition of the rock fragments can also be described using modifiers (e.g. 'basaltic', 'feldspar', etc.).

5. *MIXED SEDIMENT*
Defined as sediments composed of 40–60% siliciclastic and volcaniclastic grains and 40–60% pelagic and neritic grains.

Principal names (describes degree of consolidation).

Mixed sediment: if sediment is unconsolidated.
Mixed sedimentary rock: if sediment is lithified.

Major and minor modifiers
The mineralogy and textures of the sediment are given by major and minor modifiers. For example, a sediment consisting of 45% nannofossils, 35% quartz–sand and 20% calcareous shell debris would be termed a 'sandy nannofossil mixed sediment with bioclasts'.

Adapted and reproduced by kind permission of the authors and the Ocean Drilling Program from Mazzullo, J., Meyer, A. & Kidd, R. B. (1987). A new sediment classification scheme for the Ocean Drilling Program. *ODP Technical Note*, **8**, 19 pp.

TABLE 2 Grain Size Scale (Wentworth, 1922)

Millimeters	*Microns*	*Phi* (ϕ)	*Wentworth size class*
			Gravel
2		−1·0	
			Very coarse sand
1		0·0	
			Coarse sand
0·5	500	1·0	
			Medium sand
0·25	250	2·0	
			Fine sand
0·125	125	3·0	
			Very fine sand
0·0625	63	4·0	
			Coarse silt
0·031	31	5·0	
			Medium silt
0·0156	15·6	6·0	
			Fine silt
0·0078	7·8	7·0	
			Very fine silt
0·0039	3·9	8·0	
			Clay

TABLE 3 Classification of Chemical Sediments (Mazzullo *et al.*, 1987)

1. CARBONACEOUS SEDIMENTS

Defined as sediment composed of >50% organic material, principally plant and algal debris, that has been altered by carbonisation, putrification or bituminisation.

The two most common types of carbonaceous sediment are the coal series and sapropels.

(a) The coal series

Four ranks are recognised reflecting increasing compaction and heating:

Peat: soft, earthy organic debris in which plant material can be clearly recognised.

Brown coal: soft, dull, brown coal containing few recognisable plant fragments.

TABLE 3—*contd.*

Bituminous coal: hard black coal with bright vitreous layers, typically breaks into cuboidal fragments.
Anthracite: bright lustrous coal with conchoidal fracture.

(b) Sapropels
Sapropels are unconsolidated sludges composed of plant material (most usually algae) formed in anaerobic environments. Lithified sapropel is termed 'sapropelic coal'.

Where carbonaceous sediments contain other material such as clastic detritus. The principal names of the coal series may be modified by such terms as 'sandy', 'clayey', etc.

2. EVAPORITES
Defined as beds composed of minerals which have been precipitated from a saline solution which became concentrated through evaporation.

Evaporites are classified according to their mineralogy (e.g. halite, anhydrite, gypsum) and may be modified by terms describing their texture (e.g. 'massive', 'nodular', etc.).

3. SILICATES/CARBONATES
Chemical silicate and carbonate sediments are defined by their lack of clearly identifiable granular and biogenic components; although they may have formed through the recrystallisation of calcareous and siliceous grains. They may also form as primary precipitates (e.g. dolomite) or through authigenesis (e.g. zeolites). Such sediments are classified according to their mineralogy but may be modified by terms which describe their crystalline, as opposed to granular nature (e.g. 'crystalline', 'microcrystalline', 'amorphous', 'massive').

4. METALLIFEROUS SEDIMENTS
Metalliferous sediments are a broad category of non-biogenic, non-granular sedimentary rocks composed primarily of metal-bearing minerals such as pyrite, haematite or manganese. They are classified according to their mineralogy.

Adapted and reproduced by kind permission of the authors and the Ocean Drilling Program from Mazzullo, J., Meyer, A. & Kidd, R. B. (1987). A new sediment classification scheme for the Ocean Drilling Program. *ODP Technical Note,* **8,** 19 pp.

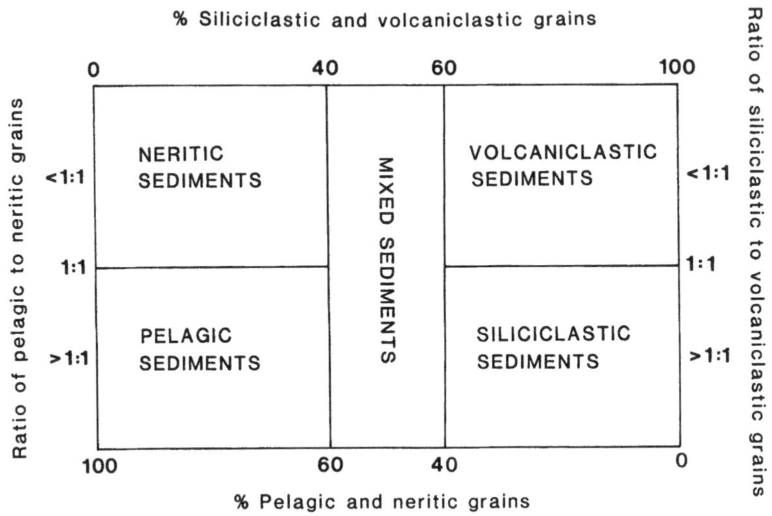

Fig. 4. Diagram showing classes of granular sediments following the classification of Mazzullo *et al.* (1987) (from Mazzullo *et al.*, 1987). (Courtesy of Dr Jim Mazzullo and the Ocean Drilling Program.)

TABLE 4 Definition of Basic Grain Types Found in Granular Sediments (Mazzullo *et al.*, 1987)

Grain type	*Definition*
Pelagic grains	Residua of planktonic microfossil tests (e.g. foraminifers, radiolarians, diatoms and nannofossils).
Neritic grains	Coarse-grained calcareous debris derived from the sublittoral zone (e.g. shell fragments, ooids) or fine-grained calcareous grains of a non-pelagic origin (e.g. micrite).
Siliciclastic grains	Minerals and lithic grains derived from the weathering of pre-existing igneous, metamorphic and sedimentary rocks.
Volcaniclastic grains	Minerals, mineraloids and rock fragments introduced into a sedimentary environment directly through submarine or terrestrial volcanism.

Sediment samples are thus described as a string of terms. In the scheme of Dean *et al.* (1985), minor modifiers are listed first and are distinguished from major modifiers by the suffix '-bearing'. When two or more minor modifiers are used, they are listed in order of increasing abundance. The major modifiers are always listed second in the string of terms and again listed in order of increasing abundance. The main name defining the sediment type is always the last name in the string. Mazzullo *et al.* (1987) also use a main name to describe the sediment class and major and minor modifiers; but differ from Dean *et al.* (1985) in placing minor modifiers after the principal name (preceded by 'with'). Thus a sediment comprising 80% fine-grained volcanic ash (mostly volcanic glass) and 20% nannofossils would be termed a 'nannofossil-bearing vitric ash' by Dean *et al.* (1985) and as a 'vitric ash with nannofossils' by Mazzullo *et al.* (1987).

Smear slide analysis therefore provides a qualitative estimate of sediment composition and is an essential first step in the description of cored sediment sequences. However, it is possible to distinguish even subtle changes in the relative proportions of different sedimentary components within a stratigraphic sequence using smear slides and these differences can be meaningful in assessing past changes in oceanic circulation, organic productivity, carbonate compensation depth, etc. (Dean *et al.*, 1985).

DISTRIBUTION OF SEDIMENT TYPES

In total, deep-sea sediments cover some 308 300 000 km^2—60% of the Earth's surface (Fig. 5). They fall into two broad groups, depending on the proportions of terrigenous and biogenic material:

(a) *Terrigenous sediments* composed largely of detrital material derived from weathering of the continents.

(b) *Pelagic sediments* which are truly oceanic in type and comprise:

- (i) Biogenic ooze composed largely of biogenic material derived from surface water productivity.
- (ii) Deep-water pelagic clay which is only found in deep areas of the ocean far from land where solution has removed the biogenic component and only the insoluble inorganic material remains.

Terrigenous sediments include turbidite muds which cover the abyssal plains; glacial material which cover substantial areas of the

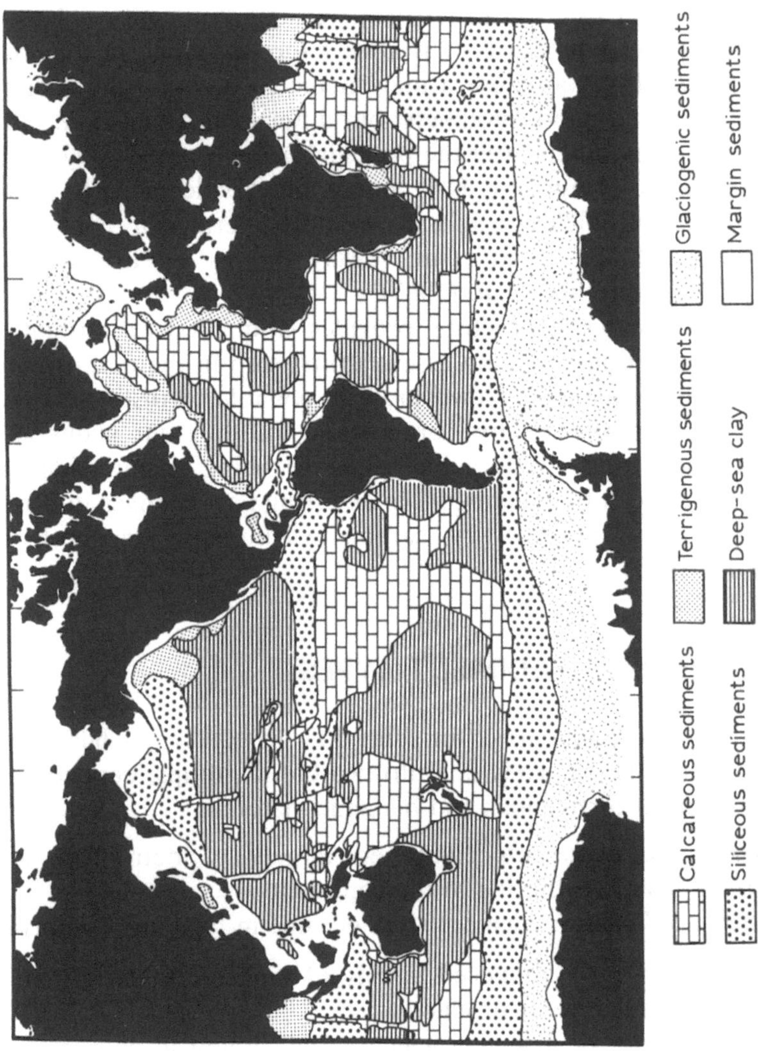

Fig. 5. Distribution of the main sediment types on the ocean floor (after Davies & Gorsline, 1976).

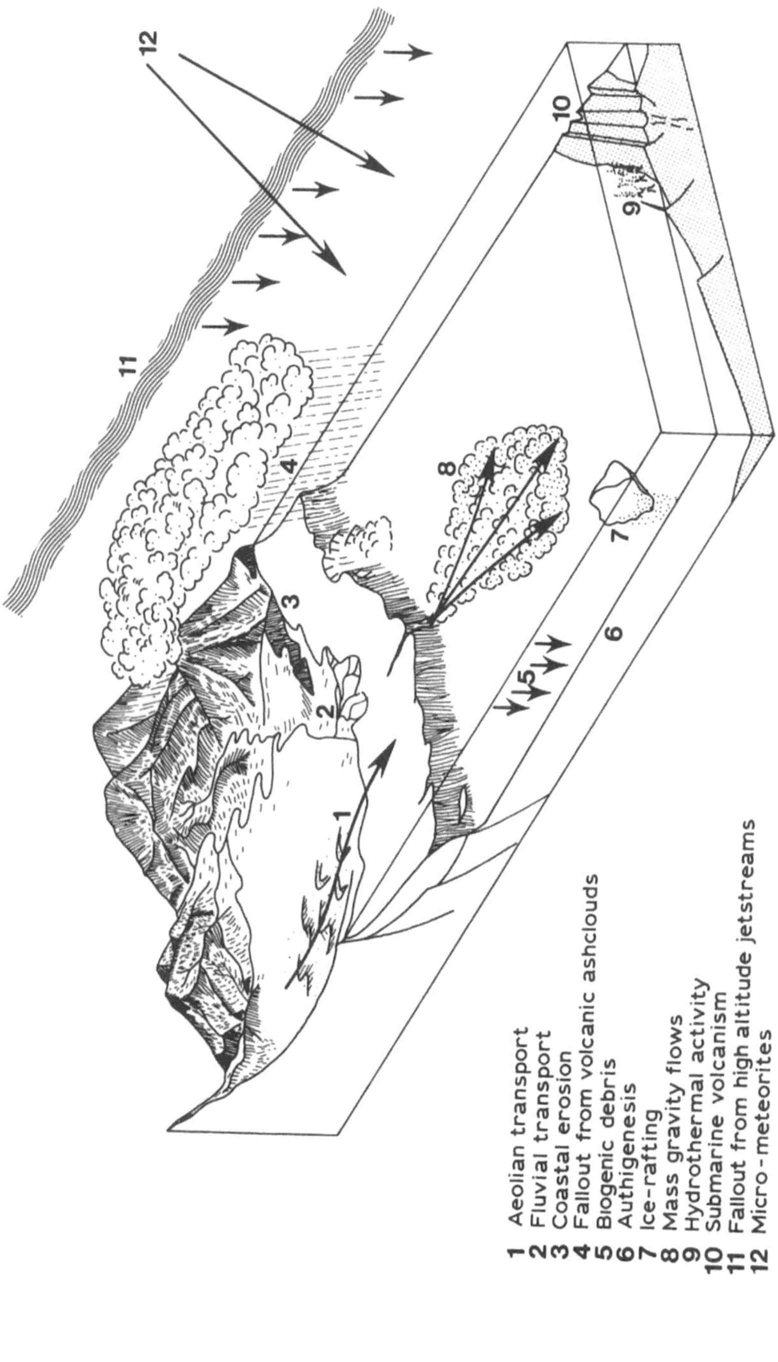

Fig. 6. The main pathways and mechanisms for the transport or formation of mineral grains in the deep sea.

seafloor in the polar regions; and slump deposits which cover areas of the continental slope and rise. Such sediments are characterised by high rates of deposition, usually small quantities of biogenic material and are usually transported into the deep ocean by some form of sediment gravity flow, for example turbidity currents or debris flows. Once in the abyssal realm, such material may be redistributed by deep geostrophic currents. These mechanisms can transport terrigenous material far into the deep ocean and far from its origin (Fig. 6). The composition of terrigenous sediments varies widely and is very dependent upon sedimentation processes together with the geology of the source area. The competence and pathways of geostrophic currents, turbidity currents and winds (in areas of high aeolian input) are very important in the distribution of detrital sediments. Besides land-derived material, terrigenous sediments usually contain some biogenic material, usually foraminifera and nannofossils and sometimes shallow-water macrofossils and plant remains. Mineralogically such sediments typically contain abundant quartz and lesser amounts of feldspar and mica. Other minerals which may be present include ferromagnesian minerals (such as pyroxenes and amphiboles), glauconite, detrital carbonate, haematite, pyrite, magnetite, chert, volcanic glass and rock fragments. Of course, the mineralogy will be strongly dependent on the source area.

Pelagic sediments include the deep-sea siliceous and calcareous oozes and red deep-water clays. The oozes are composed largely of the remains of planktonic micro-organisms such as foraminifera, coccolithophores, pteropods, diatoms and radiolaria. Their distribution is strongly depth controlled as both silica and calcium carbonate show increasing solubility with depth (Bramlette, 1961).

The calcareous oozes may either be dominated by the tests of planktonic foraminifera (foraminiferal ooze) or by the remains of planktonic plants called coccolithophores (nannofossil ooze). In either type, the other component will often be the second most important constituent. In such oozes, the terrigenous fraction (which may amount to 10–15%) will be composed mainly of quartz and clay minerals, but trace amounts of mica, chert, pyrite, iron and manganese precipitates, glauconite, rock fragments, cosmic spherules, feldspars, ferromagnesian minerals, detrital carbonate, zeolites and volcanic glass may be present. Minor biogenic components may include benthonic foraminifera, ostracods, echinoid remains, radiolaria, silicoflagellates, diatoms, sponge spicules, pteropod shells (in shallower water), fish teeth and phosphatic vertebrate remains.

The siliceous oozes are largely composed of the opaline silica tests of planktonic animals and plants. Again, there are two main varieties: diatom ooze, dominated by the siliceous skeletal remains of unicellular plants (diatoms); and radiolarian ooze, composed mainly of the siliceous tests of radiolarians. Typically siliceous ooze is only present in regions of high biological productivity (equatorial and polar belts and areas of coastal upwelling) where the depths of the seafloor is deeper than the calcite compensation depth (that depth where the rate of calcite supply equals its rate of dissolution) (Bramlette, 1961). In the North Pacific and Antarctic belts of siliceous oozes, diatoms make up as much as 95% of the bulk sediment. The mineralogical composition of the detrital fraction of siliceous oozes may be similar to that of the calcareous oozes considered above. Quartz and clay minerals again will be the dominant detrital minerals.

Deep-water pelagic clays are only found in deep areas, far from land. They usually contain less than 10% biogenic material and are mainly composed of clay minerals and fine-grained quartz, the bulk of which is derived from aeolian fall-out and has been slowly deposited from fine suspensions. Such red clays frequently contain notable amounts of authigenic minerals, such as zeolites, manganese micronodules, chert, apatite and phosphorite, as well as indicators of slow sedimentation such as fish debris and cosmic spherules. Red clays may also contain varying amounts of volcanic glass, feldspar, pyroxenes and mica.

In contrast to terrigenous sediments, pelagic sediments are characterised by low sedimentation rates, frequently contain a high proportion of authigenic minerals, extraterrestrial material and where physico-chemical conditions allow, a substantial biogenic component.

THE ABUNDANCE OF MINERALS IN MARINE SEDIMENTS

The abundance of specific minerals in marine sediments is dependent on a number of factors. The most important of these are:

(a) The availability of the mineral in the source area for detrital minerals or the availability of suitable precursors and the right physico-chemical conditions for authigenic minerals and the calcium carbonate polymorphs.

(b) The durability (i.e. resistance to mechanical breakdown) of the mineral. Absence of cleavage and a high hardness lead to a high mechanical durability.

(c) The chemical stability of the mineral. The relative chemical stability of the more common minerals is shown in Table 5. The

TABLE 5 Chemical Stability Series for Primary Minerals (after Goldich, 1938)

	Ferromagnesian minerals	Plagioclase feldspars
LEAST STABLE ↓		
	Olivine	Bytownite
	Pyroxene	Labradorite
	Amphibole	Andesine
	Biotite	Oligoclase
	Albite	
	K-feldspar (orthoclase/microcline)	
	Muscovite	
MOST STABLE	Quartz	

sequence is identical to the reaction series for high temperature petrogenesis (Bowen's reaction series). The least stable detrital mineral is olivine and the most stable is quartz. The chemical stability of a mineral is a reflection to some degree of the temperature and amount of water in the crystallising environment.

(d) The sedimentation rate. In areas of relatively rapid deposition, such as submarine fans on tectonically active margins, rapid burial will mean that even the less stable minerals may be preserved. Some authigenic minerals, e.g. ferromanganese nodules, will have their growth prevented by high sedimentation rates.

THE SMEAR SLIDE METHOD

Smear slides provide a rapid, simple and cheap technique for examining silt-size marine sediments. The very small sample size required (only 1–2 mm^3) means that even archive cores can be sampled without damage. The technique needs little preparation and simple equipment. Their usefulness for rapidly describing cored sediment sequences has been demonstrated by their continual use during the 15 years of the Deep Sea Drilling Project. Today, smear slide analysis constitutes the primary investigative tool, albeit qualitative, in the initial analysis of cored sediments and is likely to remain so. Smear slides have wide application: besides sediment cores, they can be used to examine sediment attached to dredged rocks, lithified borehole samples and unconsolidated terrestrial sediments and soils. Smear slides are especially useful in determining small scale variability in a sediment.

Sampling Strategy

During initial core description, all the major lithologies should be sampled for smear slide analysis. The number of samples taken from each core will depend on the homogeneity of the sediment (i.e. the variation in lithology seen), the presence of any unusual features or features of specific interest and any specific aims of the study. However, care must be taken when sampling layers disturbed through bioturbation or drilling disturbance to avoid sediment contaminated with younger or older material. Deep-sea oozes are often homogeneous over long distances and transitions may be gradational. Consequently, individual phases (i.e. zones of homogeneity) may be difficult to recognise.

Making a Smear Slide

To make a smear slide, a sample is taken using a toothpick or spatula (only 1–2 mm^3 is needed), after the longitudinally split core surface has been cleaned, preferably using an osmotic knife (Chemlik, 1967). The sample is then placed on a clean, clear glass microscope slide and dispersed using a drop of distilled water. The sediment/water mixture is then smeared out using a wooden toothpick. It is essential to ensure that the sediment is sufficiently dispersed to permit recognition of the finer components. The slide is then placed on a hot plate to evaporate the water, removed and allowed to cool. A few drops of a suitable mounting medium is then placed on the sample and a glass cover slip applied, care being taken to expel any air bubbles from under the cover glass. A number of mounting media are commercially available including cold setting mounting agents resistant to bubble formation such as 'Entellan', produced by E. Merck, D-1600 Darmstadt, FRG and marketed in the United Kingdom by BDH Chemicals Ltd. At room temperature, 'Entellan' will set in 20 minutes but will not set hard for up to two days. When applying the cover slip, care must be taken to try to ensure that the cover slip is sealed right up to each edge. Excess embedding agent can be removed with a cloth or tissue moistened with acetone. To ensure operator comfort, the area where the smear slides are made should be well-ventilated; 'Entellan' contains xylene and is harmful when inhaled (it is also inflammable). If other mounting media are used, operators must acquaint themselves with any toxic hazard involved.

Smear slides can only really be successfully made if the sediment is in the silt range (4–63 µm). With coarser sediments, it will be difficult to avoid air bubbles forming however, the operator may have some success if the embedding agent is placed on one end of the slide and allowed to slowly run over the sample. The cover slip is then carefully applied.

The Examination of Smear Slides

The examination of smear slides requires a good quality petrological microscope, preferably fitted with a mechanical stage. The components of the sediment are identified and their relative volume percentage estimated. To ensure consistency in percentage estimates, a percentage of smear slide analyses should be duplicated (at least 40%). Further, quantitative determination of carbonate percentages should provide some control on visual estimates.

Both Folk (1951) and Shvetsov (1954) have demonstrated that in purely subjective assessments of volume percentages there is a strong tendency to over-estimate, especially if the component is only present in small amounts. To remedy this, both Folk (1951) and Shvetsov

(1954) (reproduced in Terry & Chilingar, 1955) have produced a series of comparative percentage diagrams as guides. These were produced by cutting out two circles of equal diameter, one white and one black and then cutting up a section of the black card equal in area to the desired percentage and sticking the pieces to the white circle. A series of percentage charts, made in this way, is included as Appendix 2. Such charts are of great value and wide application, being useful for smear slides, thin sections, hand specimens and even aerial or bottom photographs showing pebble or outcrop cover. Allen (1956) has criticised such percentage charts by pointing out that contrasted particles may seem larger or smaller (and hence percentages may be over or under-estimated) depending on the background colour. Thus, the use of neutral grey instead of strongly contrasting black or white usually results in visual estimates that differ from estimates made on the contrasting sets. Ideally, a series of monochrome and polychrome percentage charts should be constructed to enable petrographers to recognise the psychological illusions. In fact, Allen (1956) reports that, using such a series of charts, students were able to estimate consistently percentages within 5% of modal analyses made with recording micrometer stages.

Even with the use of comparative percentage charts, there may be some operator variation in estimating the proportions of minerals in smear slides. In a wide-ranging study of operator variation, Griffiths & Rosenfeld (1954) have shown that experience does not necessarily result in the removal of operator differences but tends to make individual operators more consistent. If anything, they conclude, experience tends to increase operator bias and to increase differences between operators.

Despite the criticisms of comparative percentage charts, the potential errors are probably of little significance, as the usefulness of smear slide analysis lies mainly in the recognition of the relative importance of the various components making up the sediment. The ultimate test of the method is whether it compares favourably with results from quantitative analytical techniques. During the Deep Sea Drilling Project, sediments were named according to the results of shipboard smear slide analyses and only rarely did primary names, given using a particular sediment classification, need altering as more information came available through shore-based X-ray mineralogy or grain size analyses (Kidd, 1975). These later studies changed/or added 'modifiers' to the primary name. Further, smear slide analysis has the overwhelming advantage over quantitative X-ray techniques of allowing chemically similar but distinct components to be distinguished. Examples are provided by crystalline quartz and volcanic glass; and crystalline carbonate minerals and biogenic carbonate debris.

In summary, smear slides provide a most effective technique for initial analysis of sedimentary components in marine sediments. It remains unrivalled as a cheap, simple and rapid investigative method.

PART II

MINERALS AND MINERALOIDS OCCURRING IN MARINE SEDIMENTS

MINERALS AND MINERALOIDS OCCURRING IN MARINE SEDIMENTS

EXPLANATORY NOTES

All the major entries in this section are in alphabetical order. Olivines, pyroxenes and amphiboles are grouped together under 'ferromagnesian minerals'. Heavy minerals are also grouped together. At the beginning of each mineral description, the chemical composition and formula are given together with an indication of the likely origin of the mineral (e.g. detrital, biogenic, volcanogenic or authigenic). The optical properties, stability, occurrence and oceanic distribution of each mineral or mineraloid are then summarised as are its main distinguishing features. A short note is included on its appearance in thin section. Within this framework, the following parameters have been selected for discussion.

Form/shape. This is often one of the most distinctive properties for identification. Many of the minerals seen in marine sediments have distinctive and consistent forms which allow easy identification. Aragonite needles, dolomite rhombs, tabular feldspar (when euhedral or subhedral), flaky micas, cuspate and lunate glass shards, phillipsite prisms are all characteristically distinctive in form and easily recognised. As the minerals in a smear slide will be unconstrained, minerals in which particular forms dominate will usually settle in preferred orientations and this will mean that certain optical characters and characteristics, often important in thin section identification, will not be seen in smear slides. Flaky minerals like micas will tend to settle with {001} parallel to the plane of the slide. Prismatic crystals, like pyroxenes and amphiboles, tend to settle so that their long axes are parallel to the plane of the slide. Therefore micas will tend to be seen as flat ragged plates and basal sections of euhedral or subhedral amphiboles and pyroxenes will be rarely seen.

Fragmentation characteristics. Minerals with well-developed cleavage often tend to part along these planes during transport and abrasion in the sedimentary cycle. Thus cleavage fragments of feldspar, amphiboles, pyroxenes, gypsum and mica may be common in some sediments. Minerals of delicate habit such as phillipsite twins rarely survive the making of the smear slide without fragmentation.

Grain-size range. Almost all the minerals which can be identified in smear slides will be in the silt–sand size range. Within this, most minerals can show a wide variation in size depending on their residence time in the sedimentary cycle and original size. Some minerals are characteristically small when they occur. Examples are aragonite needles (commonly 10 μm or smaller), clays, dolomite rhombs (frequently 15 μm or smaller), anatase and monazite (10 μm) and clinoptilolite (commonly under 30 μm). Micas with their flaky habit have a hydraulic equivalence of much smaller particles, so may be conspicuously large when compared with the associated material.

Transparency/colour in plane-polarised light (PPL). These can be important factors in identification. Aragonite, calcite, dolomite, olivine, some amphiboles and pyroxenes, apatite, monazite, zircon, gypsum, sanidine, muscovite, quartz, some volcanic glass, phillipsite and clinoptilolite are all transparent and colourless. Feldspars (except sanidine) are also colourless but may be turbid due to alteration. Minerals with distinctive colour are glauconite (vivid green), biotite mica (brown or green) and translucent iron oxides (red or brown). Amphiboles are commonly coloured and the pyroxene aegirine–augite is a distinctive inky blue-green. Palagonite is usually a characteristic golden yellow-brown. Pyrite, micronodules and cosmic spherules are always opaque.

Relief. The relief shown by a mineral (i.e. its prominence in the field of view) depends on the difference between the refractive index of the mineral and that of the mounting medium. The greater the difference, the greater the relief. If the refractive index of the mineral is close to that of the mounting medium then the grain will hardly be visible. Minerals with low refractive indices and those with high refractive indices will show high relief. In most standard texts, mineral refractive indices are usually compared to Canada Balsam but the demise of balsam as a common embedding agent for microscopy means that refractive indices are probably best compared to quartz ($n_\omega = 1{\cdot}544$, $n_\varepsilon = 1{\cdot}553$). Minerals which show diagnostically high relief compared to quartz are monazite, sphene, zircon, anatase, rutile, garnet and epidote. Some minerals show diagnostically very low relief, for example, calcite in one direction and indeed this property can be used to distinguish it from dolomite in rhombohedral form.

Differences in relief can be determined optically by the Becke line test which can be used with mounting media of any refractive index. When the illumination of a smear slide or thin section being viewed in plane-polarised light is reduced by means of the substage diaphragm, a bright line of increased illumination (the Becke line) will be seen at the boundaries of grains that differ in refractive index to the mounting medium. The Becke line is best found when the edge of a suitable crystal is viewed under medium power and the substage diaphragm closed. The cause of this bright line of illumination is that, in nearly all cases, the grain surface is inclined and therefore light entering the slide from below will be transmitted both through the grain and the mounting medium. At the bounding edge both reflection and refraction will occur. As a result, a portion of the entering beam is refracted one way or the other depending on the relative refractive indices of the grain and the mounting medium. If the refractive index of the grain is greater, light will strike the inclined grain boundary in some places at angles greater than the critical angle. Part of the beam will then be deflected towards the grain, resulting in the light blur termed the Becke line. If the microscope's focusing control is now rotated in such a way that the distance between the stage and objective increases, the Becke line will move into the phase having the higher refractive index. Thus the refractive index of grains may be compared to the mounting medium whose refractive index can be precisely known. The refractive index of loose grains may be determined by the use of 'standard liquids'—an incremental series of liquids whose refractive indices are precisely known. Immersion of the grain in successive standard liquids followed by a Becke line test will enable its refractive index to be determined. All refractive index values given in the text are taken from Kerr (1977).

General values for refractive index are designated by the letter n. Isotropic minerals, having the same optical properties in all directions, will only have one refractive index; but in anisotropic minerals refractive index will vary with direction. With uniaxial minerals (i.e. those crystallising in the hexagonal and tetragonal systems) two refractive indices are normally given, that for the ordinary ray n_ω and that for the extraordinary ray, n_ε. In optically positive uniaxial crystals, n_ω will be less than n_ε; while in optically negative uniaxial crystals, n_ω

will be greater than n_ε. With biaxial minerals (orthorhombic, monoclinic and triclinic crystals) three refractive indices are usually given n_α, n_β and n_γ. The greatest refractive index (n_γ) is associated with the slow ray and the least refractive index (n_α) with the fast ray. n_β is an intermediate value in the direction at right angles to the other two. n_β is sometimes called the mean refractive index but it is not the arithmetic mean of the other two and should not be thought of as such. Intermediate refractive index is a better term.

Birefringence. With the exception of amorphous minerals and those crystallising in the cubic system, light passing through minerals will split into two beams, travelling at different speeds, corresponding to two different refractive indices. The difference between the greatest and the least refractive index of a mineral gives the numerical value of birefringence which, together with grain thickness, determines the interference colours shown by the mineral when seen in cross-polarised light. The wide variation in thickness shown by grains in smear slides makes interference colours a less reliable guide to the identification of minerals than in conventional thin sections, but as 30 μm is the coarse/medium silt boundary, quartz and feldspars, the commonest detrital minerals, usually still show interference colours within the first order. Although interference colours are not absolutely reliable, birefringence is still a useful guide. Minerals with a high birefringence, such as aragonite, calcite and dolomite, will exhibit bright pastel shades of high order yellows or whites which are distinctive. Weakly birefringent minerals, such as quartz, feldspar, gypsum, phillipsite and clinoptilolite, will show low order interference colours. However, there are a few anomalies; the micas, for example, have strong birefringence but as they take up a preferred orientation in smear slides with {001} usually approximately parallel to the plane of the slide, they will present basal sections to the viewer and these have low birefringence.

Alteration. Many minerals are unstable under oceanic conditions as are volcanic glasses and thus may show alteration rims, turbidity or may be pseudomorphed. Aragonite, feldspar, olivine, biotite and palagonite are all unstable in sedimentary environments and will be associated with alteration products. Some minerals, notably zeolites and palagonite, are themselves formed through the alteration of volcanogenic material.

Crystal system. The crystal system to which a mineral belongs will affect its form and control some of its optical properties.

Cleavage. Many crystals have the property of splitting along well-defined cleavage planes, parallel to actual or possible crystal faces. Cleavage will greatly affect the fragmentation characteristics of the mineral and lessen its durability. Cleavage is an important property allowing extinction angles to be measured/estimated.

Pleochroism. Certain coloured minerals show a variation in colour as the microscope stage is rotated in plane-polarised light. This property is called pleochroism. Mineral colour in smear slides results from the absorption of certain wavelengths from the incident white light, the resulting transmitted light being complementary in colour to that absorbed. Pleochroic minerals absorb different wavelengths in different directions. Pleochroism is an important diagnostic property. Many amphiboles are strongly pleochroic but the only pyroxenes that are, are hypersthene, aegirine and aegirine–augite. Biotite mica is pleochroic but this will rarely be observed in smear slides as basal sections of biotite are dark coloured and show little, if any, pleochroism as the stage is rotated.

Extinction angle. When viewed in cross-polarised light and the stage is rotated, anisotropic grains will become dark in four positions that are 90° apart. Once dark, the grain is said to be at extinction. Extinction occurs whenever either of the vibration directions of the grain becomes parallel with the vibration direction of the polariser. In this position, all the light transmitted through the grain utilises that which is parallel to the polariser vibration direction. If these rays are cut out by the analyser set with its vibration direction at right angles to that of the grain, then the grain will appear dark. Frequently, mineral grains will have cleavage lines or crystal boundaries that allow the angle at which extinction occurs to be referred to a crystallographic feature. Of course, with anhedral grains, extinction angles will be indeterminate. If extinction occurs when the cleavage is parallel to the vibration direction of the polars (i.e. N–S, E–W), extinction is said to be straight or parallel, but if extinction occurs when cleavages or crystal outlines lie at inclined angles to the vibration directions of the two polars, extinction is said to be oblique. In some minerals, e.g. dolomite, the cleavage patterns or crystal outlines may have rhombic cross-sections. In many cases, the crystals extinguish when the vibration directions of the polars fall parallel to the rhombic diagonals. Such extinction is termed symmetrical. Measurement of extinction angles is important in distinguishing between ortho and clinopyroxenes and between clinopyroxenes and amphiboles. The different varieties of plagioclase feldspar are also distinguished on extinction angles.

Twinning. Crystal twinning can be an important diagnostic feature in the identification of some minerals grains. Twinning in feldspar is very obvious in cross-polars and the type of twinning is diagnostic of variety. Cruciform twinning is common in phillipsite.

Optical character. The relationship between the refractive indices of the mineral and its crystallographic symmetry can be visualised by drawing in all directions from the centre of a crystal lines whose lengths are proportional to the refractive index for the particular vibration direction. The resulting geometrical figure is termed the indicatrix. The lines will form the axes of the ellipsoid and are commonly termed X, Y and Z. Usually X denotes the 'fast ray' and Z the 'slow ray', with Y for the intermediate ray perpendicular to the X–Z plane. The corresponding refractive indices are designated as α, γ and β respectively. In cubic minerals and non-crystalline substances, the refractive index is the same in all directions (i.e. $X = Y = Z$), thus the form of the indicatrix is a sphere and the mineral is isotropic. With minerals crystallising in the hexagonal and tetragonal systems the indicatrix has the form of a rotation ellipsoid in which all sections perpendicular to the c axis of the crystal are circular (i.e. $X = Y < Z$ or $X < Y = Z$). Such minerals are termed uniaxial and have only one optic axis. Minerals crystallising in the orthorhombic, monoclinic and triclinic systems have an indicatrix of lower symmetry—a triaxial ellipsoid, corresponding with the lower crystallographic symmetry. In such an ellipsoid, the three axes are unequal ($X < Y < Z$) and such crystals have two optic axes and three indices of refraction and are termed biaxial. Both uniaxial and biaxial minerals are anisotropic and give different interference figures when thin sections of crystals, cut perpendicular to the optic axis (uniaxial case) or to the acute bisectrix (biaxial case) are viewed under cross-polars using convergent light. Crystals in smear slides will rarely be well presented for this kind of work although occasionally, grains may be sufficiently well orientated to yield off-centre figures which may be capable of providing partial data in some cases. The reader is referred to one of the excellent standard mineralogical texts such as Kerr (1977) for more detail on crystal optics.

Group variability/related species. Feldspars, olivines, pyroxenes and amphiboles all form solid solution series. Consequently, there may be variation in their optical properties which can be used to distinguish the members of the series. In the feldspars, pyroxenes and amphiboles, individual members may be identified using the maximum extinction angles in longitudinal sections of euhedral crystals (Kerr, 1977) and this can be used in smear slides providing the crystals show good crystal form. Quartz may show a variety of morphotypes dependent on its ultimate origin (Krynine, 1946; modified by Folk, 1980; Basu 1985) and in local instances it may be possible to make specific observations on quartz provenance although most of the quartz seen in marine sediments will be angular, anhedral and faintly strained or unstrained.

Associated minerals. Some minerals may occur in characteristic associations reflecting the similar environmental conditions required for their formation/preservation. Manganese nodules are frequently associated with fish debris and altered volcanic rock fragments, reflecting the low sedimentation rates required for their formation and the necessary availability of iron and manganese for oxidation and precipitation (derived through the weathering of volcanics). Sometimes micronodules may occur in association with rhodochrosite, as the two forms are end members of a continuous transformation series, with rhodochrosite transforming to manganese oxide grains with increasing redox potential (Borella & Adelseck, 1979). Palagonite, volcanic glass, pyroxenes and zeolites are frequencly associated, reflecting their volcanic derivation.

Occurrence. The occurrence of minerals in marine sediments is dependant on several factors which will vary depending on the mineral concerned. Availability, durability, stability, sedimentation rate and the right physico-chemical conditions are probably the most important. The conditions affecting the occurrence of the individual minerals are discussed under this heading and factors which may restrict their occurrence are noted.

Oceanic distribution. Over the last thirty years, there has been much research involving mineralogical analyses of marine sediments. As a result, variations in the oceanic distribution of many minerals are now fairly well known. Variation in the proportion of detrital minerals in the sediments of different oceans may relate to variation in the ratios of the land drainage areas to those of the specific ocean basin. Minerals largely of continental derivation such as mica and quartz will be most common in shelf, slope and rise sediments. The oceanic distribution of the calcium carbonate polymorphs will be dependent on depth (Bramlette, 1961). With some minerals, such as the feldspars, the different types may form distinct provinces depending on the geotectonic setting (Heath, 1969).

Origin/source. Under this heading, the ultimate source of mineral types is discussed. Many minerals can indicate a specific provenance and hence can be extremely useful in many geologic studies. Some minerals such as apatite, sphene, zircon, rutile, kyanite and garnet are, when they occur, exceptionally useful in this way (Pettijohn, 1975). Certain morphotypes of quartz may also be useful as may be types of feldspar (see Folk, 1980; Pettijohn, 1975).

Transport mechanisms. The pathways and mechanisms for the transport of minerals into and within the deep sea are many and varied. Aeolian transport, fluvial transport, coastal erosion, ice-rafting, mass gravity flows and geostrophic currents are probably some of the most important. Aeolian transport is particularly important for fine-grained material (Windom, 1969). Some transport mechanisms may have particular relevance for particular minerals or mineraloids, e.g. atmospheric fall-out from volcanic ashclouds for volcanic glass shards.

Diagenesis. Many minerals are unstable in the marine environment and readily alter. Some of the minerals and mineraloids listed in this section are precursors to others and may be commonly associated with their alteration products. For instance, basaltic glass hydrates to yellow palagonite and the constituents lost from the glass during the palagonitisation process are precipitated as zeolites and calcite. Thus, where relevant, aspects of diagensis are also discussed.

Downhole variability. Some minerals may only be stable under near-surface conditions and show progressive dissolution with depth. Therefore downhole variability must be considered, especially when examining deep cores. Dissolution with depth is seen in phillipsite and indicated by etched crystal faces and decreasing frequency of occurrence. Indeed downhole, phillipsite is increasingly replaced by clinoptilolite (Stonecipher, 1976) which itself is rarer in surface and shallowly buried sediments (Kastner & Stonecipher, 1978). Ferromagnesian minerals, especially pyroxenes and amphiboles, may show downhole etching due to intrastratal solution as will many 'heavy' minerals. Such grains will be corroded and pitted and may show characteristic 'hacksaw' terminations.

Indicator. The significance of each mineral or mineraloid in a marine sediment is summarised under this heading. As certain minerals (e.g. pyrite, zeolites) require specific physico-chemical conditions for their formation, their presence in a sediment will be very informative concerning water depths, redox potentials, bottom conditions, etc. Other minerals indicate low sedimentation rates (e.g. manganese nodules and zeolites) or basaltic volcanism (e.g. zeolites, palagonite and pyroxenes). Certain detrital minerals (e.g. many heavy minerals) are very useful provenance indicators (Pettijohn, 1975). Feldspars can in some instances have palaeoclimatic usefulness (Folk, 1980). This section is probably best consulted once a mineral or mineraloid has been identified from its optical properties as it will guide further interpretation.

Main distinguishing features. This section summarises or indicates the main diagnostic criteria for identifying the minerals as seen in smear slides and provides a quick reference guide to identification.

Appearance as loose grains. This summarises the appearance of the minerals when seen in reflected light. It should be useful when examining sand-sized material using a binocular microscope.

Appearance in thin section. This summarises the appearance of the minerals when seen in conventional thin sections (at a thickness of 30 μm).

Notes/remarks. This final section includes any other relevant information regarded as useful to the user.

Of course, duplication and occasional inapplicability will mean not all these parameters are included for each mineral or mineraloid described in this section, but the basic scheme is followed for each of the major entries.

All sections are illustrated using photomicrographs. Captions describe the mineral shown, the age of the sample and its origin. Where known, the latitude and longitude of the core, from which the sample is taken, is also given. Samples from Deep Sea Drilling Project (DSDP) legs have sample numbers consisting of a string of three numbers separated by hyphens. The first number indicates the site number, the middle number gives the core number and the last number indicates the section number. The number immediately following this string gives the depth in the core section, from which the sample was taken, in centimetres.

ARAGONITE

Biogenic/detrital/authigenic
$CaCO_3$
Calcium carbonate

Form/Shape

Aragonite may occur as fine or stubby acicular needles, sometimes resembling grains of rice or wheat (Figs 7–9); chisel-shaped short prismatic crystals or, when biogenic, spired or conical shells and shell fragments (Fig. 10).

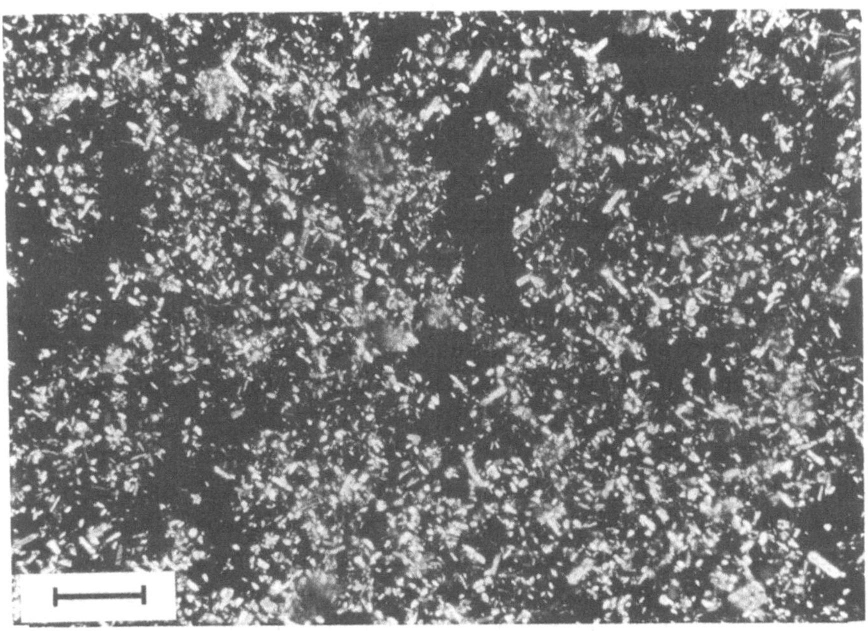

Fig. 7. Low magnification view of aragonite needles seen in cross-polarised light. Although fine-grained, the acicular form is clearly visible. Lower Pleistocene, Black Sea (42°05.94′N, 29°36.82′E). DSDP Leg 42, 380A-60-3, 1 cm. Scale bar equals 0·05 mm.

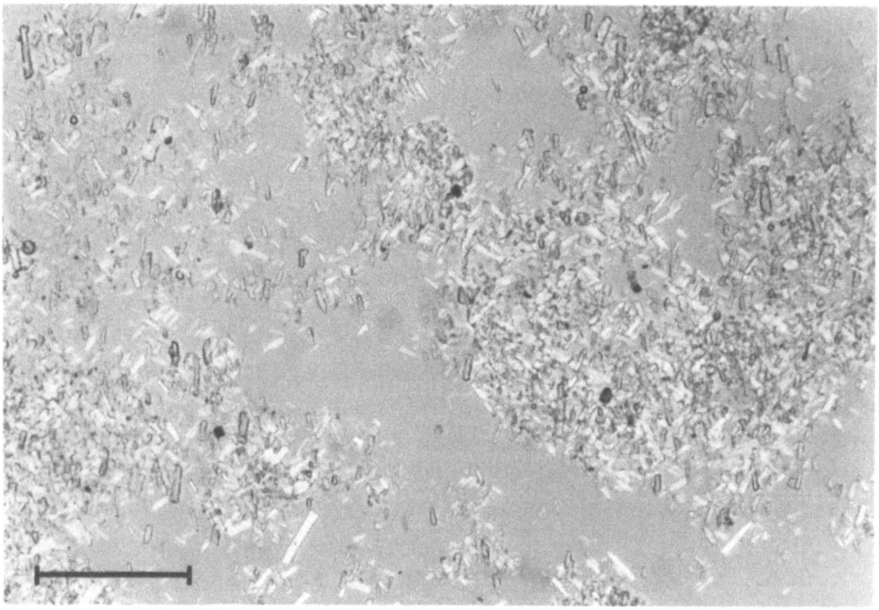

Fig. 8. Aragonite needles. Lower Pleistocene, Black Sea (42°05.94′N, 29°36.82′E). DSDP Leg 42, 380A-63-4, 4 cm. Plane-polarised light, scale bar equals 0·05 mm.

Grain Size Range

Variable. Aragonite needles are commonly 10 μm or smaller. Aragonite crystals usually lie in the silt-size range. Organic aragonite fragments may be sand-size or larger.

Transparency/Colour in PPL

Usually colourless and transparent. Biogenic aragonite may show some form of surface sculpture such as growth lines.

Relief

Refractive indices: $n_\alpha = 1{\cdot}530$, $n_\beta = 1{\cdot}682$, $n_\gamma = 1{\cdot}686$. Thus like calcite, relief varies with direction, although this may not be obvious.

Birefringence

Extreme (0·156). Grains will show interference colours of high order most frequently in pastel shades of yellow, grey or white.

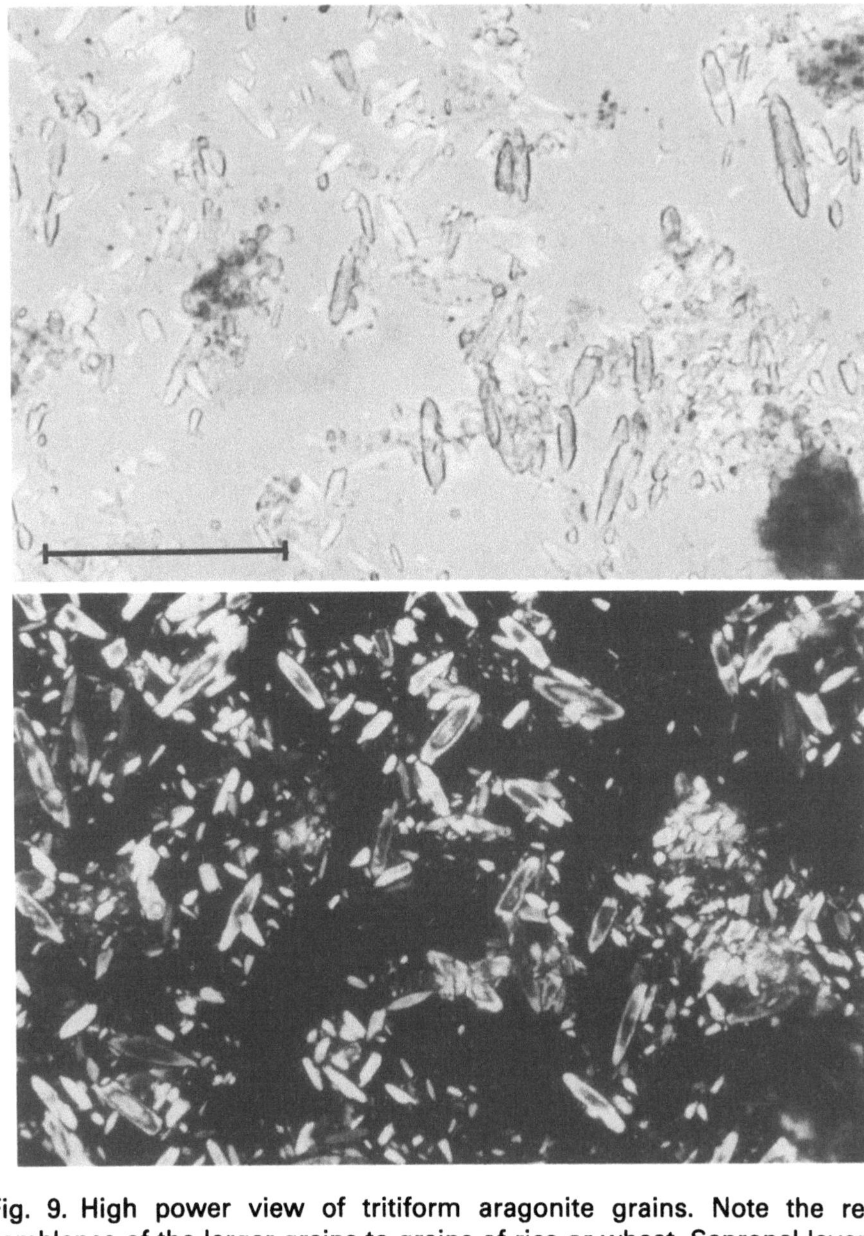

Fig. 9. High power view of tritiform aragonite grains. Note the resemblance of the larger grains to grains of rice or wheat. Sapropel layer, Pleistocene, Black Sea (43°00.29′N, 36°00.68′E). DSDP Leg 42, 379A-11-5, 47 cm. Viewed in plane-polarised light (top) and cross-polarised light (bottom). Scale bar equals 0·05 mm.

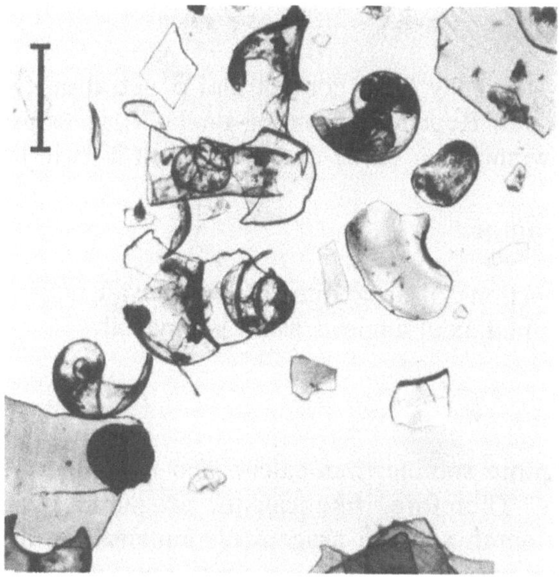

Fig. 10. Biogenic aragonite (spired and planispiral shells and shell fragments). Pteropod sand, Pleistocene, northeast of Atlantis II Deep, Red Sea. Plane-polarised light. Scale bar equals 0·5 mm.

Alteration

Aragonite is metastable under oceanic conditions and inverts, over time, to calcite.

Crystal System

Orthorhombic.

Cleavage

Distinct parallel to the length of the crystals {010}. {110} poor.

Pleochroism

Absent

Extinction Angle

Straight extinction in prismatic grains (inorganically precipitated aragonite).

Twinning

Aragonite crystals may form contact and penetration twins with {110} as the twin plane. Repeated twinning on {110} gives pseudohexagonal aggregates. Twinning on {110} may also result in twin lamellae.

Optical character

Basal {001} sections of aragonite show a biaxial negative interference figure with a small axial angle. Calcite is uniaxial.

Occurrence

Aragonite is more soluble than calcite and is highly undersaturated in the deep sea. Therefore, like calcite, aragonite deposition will be controlled by depth and the aragonite compensation depth (the depth contour which separates aragonitic from aragonite-free sediments) is considered by Morse & Berner (1979) to be generally 3 km shallower than the calcite compensation depth (CCD).

Aragonite needles (up to a few microns long) are a major component of shallow-water carbonate muds in sub-tropical and tropical regions. Some are derived from the decomposition of calcareous algae, the rest result from the direct precipitation of aragonite. The solubility of calcium carbonate in seawater is controlled by temperature and the amount of dissolved carbon dioxide. Microscopic plants cause the direct precipitation of aragonite needles in that they abstract carbon dioxide from seawater during photosynthesis thus lowering the solubility product for calcium carbonate. Such needles may be carried into the deep sea by turbidity currents and other mass gravity flows where rapid burial may inhibit dissolution (Jipa & Kidd, 1974).

Aragonite shells are secreted by some groups of planktonic organisms of which pteropods—a group of opistobranch gastropods—are the most important (Fig. 10). Pteropods are widespread in the world ocean and may constitute a high percentage of calcareous oozes above the aragonite compensation depth (ACD).

Aragonite may also form from hydrothermal systems at mid-ocean ridges, portions of which lie above the ACD in both the Atlantic and Indian Oceans. Thompson *et al.* (1968) have described aragonite nodules recovered during dredging on the Mid-Atlantic Ridge (43°N), the Romanche Fracture Zone and around St. Paul's Rocks in the equatorial Atlantic. Kharin (1974) has also described reniform nodules of aragonite on weathered surfaces of serpentinised peridotites from the rift zone of the Mid-Atlantic Ridge. These nodules comprised aggregates of radiating aragonite needles and prismatic crystals. Locally, aragonite crystals were abundant in black serpentinite sands. From the Arabian–Indian Ridge in the Indian Ocean, Rozanova (1971) has reported columnar growths of aragonite crystals in the cement of tectonic breccias. In all of these cases the formation of the aragonite relates to the release and reprecipitation of calcium oxide during the serpentinisation of peridotites (Miyashiro *et al.*, 1969).

Oceanic Distribution

Pteropod ooze is considered to cover only 0·6% of the ocean floor (Sverdrup *et al.*, 1942; Berger, 1976) but occurs in relatively significant quantities in the Atlantic Ocean (on the flanks of the Mid-Atlantic Ridge) (Sverdrup *et al.*, 1942). Pteropods also appear abundant in Recent sediments in the Mediterranean, Red Sea and Northwest Indian Ocean (Herman, 1978). Otherwise, aragonite is relatively unimportant in deep-sea sediments.

Main Distinguishing Features

The best guide to the identification of aragonite is its form, especially when it occurs as needles. However, in general optical properties, aragonite resembles calcite but lacks the rhombohedral cleavage and has a higher refractive index. Further, aragonite is biaxial, whereas calcite is uniaxial.

Appearance in Thin Section

Plane-polarised light: consolidated shell debris or colourless columnar or fibrous aggregates. Relief may vary with the orientation of the columns of fibres. Cross-polarised light: interference colours are high order whites. Extinction parallel to crystals or columns.

BARITE

Authigenic/detrital (rare)
Barium sulphate
$BaSO_4$

Form/Shape

In marine sediments, barite occurs as prismatic, euhedral to angular, fractured, anhedral crystals and crystallites (Figs 11 and 12). It also occurs as sharply angular cleavage fragments, microcrystalline aggregates and more rarely as delicate fans and double-ended feathery 'dumb-bells'. Crystals are often flattened on {001} and these are easily distinguished from drilling mud which is quite anhedral. (Barite is used in oil and gas well drilling as 'heavy mud' to support drill rods and to prevent gas blow-outs.) Detrital barite is rare in marine sediments.

Grain Size Range

Barite crystals may reach up to 60 μm, but are usually no larger than 10 μm. Large crystals are rare, although large crystals have been reported associated with ferromanganese deposits (Church, 1970).

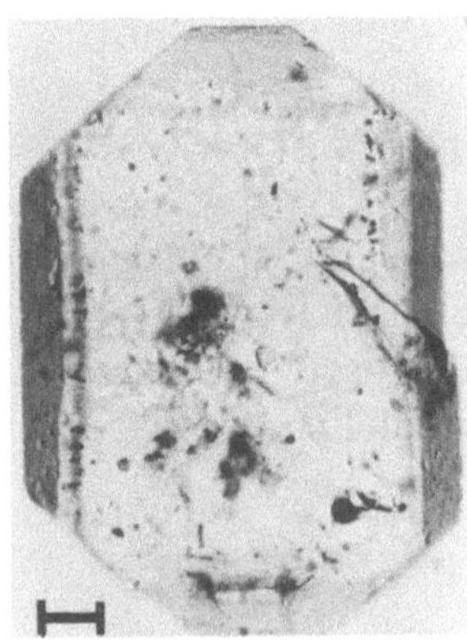

Fig. 11. Euhedral tabular barite crystal. Upper Cretaceous, Canary Basin, Northeast Atlantic (25°55.53′N, 27°03.64′W). DSDP Leg 14, 137-6-1, 127 cm. Plane-polarised light, scale bar equals 0·05 mm. (Courtesy of Dr W. H. Berger and the Deep Sea Drilling Project.)

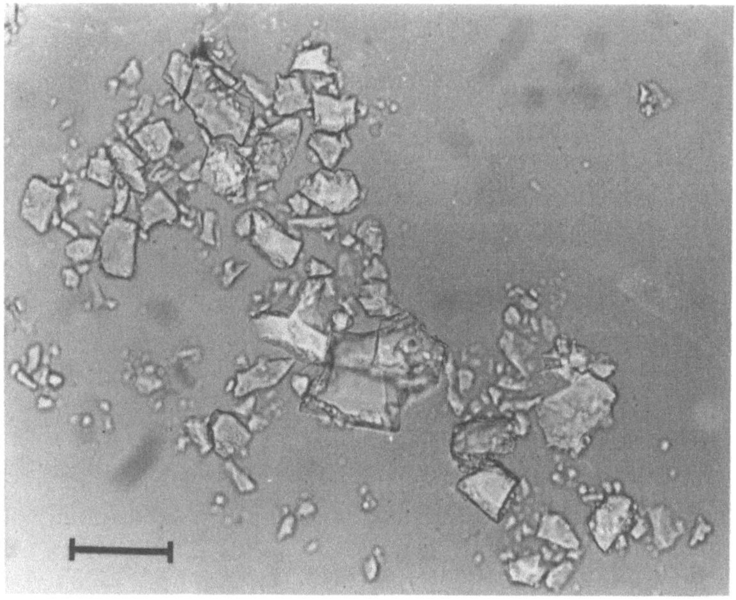

Fig. 12. Anhedral barite. Middle Miocene, Coiba Ridge, Eastern Equatorial Pacific (06°07.38'N, 81°02.62'W). DSDP Leg 16, 155-6-4, 105 cm. Plane-polarised light, scale bar equals 0·05 mm.

Transparency/Colour in PPL

Transparent to translucent, usually colourless, although thick grains may be yellowish in colour. May sometimes appear turbid due to inclusions.

Relief

Moderate–high. Refractive indices: $n_\alpha = 1{\cdot}636$, $n_\beta = 1{\cdot}637$, $n_\gamma = 1{\cdot}648$.

Birefringence

Rather weak (0·012), therefore interference colours will tend to be low order.

Crystal System

Orthorhombic.

Cleavage

Perfect {001} and {110}, imperfect {010}.

Pleochroism

Absent.

Extinction Angle

Prismatic grains show straight extinction (parallel to the {001} cleavage).

Twinning

Polysynthetic twinning with {110} as the twin plane occasionally occurs.

Optical Character

Biaxial, optically positive.

Occurrence

Barite is fairly widespread in marine sediments, although generally it averages less than 1% of most sediments. It is frequently associated with biogenic material and occasionally occurs within burrows. Barite is relatively common in sediments beneath areas of high organic productivity (Arrhenius & Bonatti, 1965; Church, 1970).

High concentrations of barite (up to 10% of the carbonate-free fraction) have been reported from the sediments of the East Pacific Rise (Arrhenius & Bonatti, 1965; Elderfield, 1976). In this case, its origin may relate to hydrothermal activity at the crest of the rise (Arrhenius & Bonatti, 1965; Bostrom *et al.*, 1971). However, barite appears relatively rare in the North Pacific, although Revelle & Emery (1951) describe barite concretions found off the coast of California and ascribe them to local hydrothermal activity on the seafloor. However, Goldberg *et al.* (1969) have shown that although these nodules may have a hydrothermal origin, they are not *in situ* and may have been transported from the adjacent land mass, perhaps forming in restricted lagoons. Barite concretions have also been reported from the seafloor off California and Indonesia by Sverdrup *et al.* (1942).

Dean & Schreiber (1977) have reviewed the occurrence of barite in sediments cored by the Deep Sea Drilling Project. They found that barite is far more abundant in cores from the eastern tropical Pacific than from anywhere else so far cored. This supports Cronan's (1974) observation that subsurface and surface distributions of barite are very similar. They further reported that although barite occurs in sediments of all ages, it appears more abundant in older sediments in the Atlantic and Indian Oceans than in the Pacific. They also noted that generally barite tends to increase in abundance with depth within a core at a given site. Although Church (1970) and Cronan (1974) emphasise the association of barite with calcareous sediments, Dean & Schreiber (1977) found no universal correlation with sediment type.

Origin

Both a hydrothermal origin and a geochemical origin, influenced, at least in part, by biological factors have been proposed for the formation of deep-sea barite. Both proposals seem likely and can explain most observed associations. A hydrothermal origin for barite concretions found off California has been proposed by Revelle & Emery (1951). Arrhenius & Bonatti (1965) also propose a hydrothermal origin for the high concentration of barite found in pelagic sediments of the East Pacific Rise. They propose that barium entering seawater from volcanic sources near the ridge crest reacted with dissolved sulphate to precipitate barite. However, the association of barite with biogenic debris and organic matter is well-documented and Church (1970) emphasises the importance of biological activity in its formation. Barium is concentrated by a number of marine organisms (Bowen, 1956; also see Tables 2 and 3 in Gooday & Nott, 1982). Relatively enormous quantities of barium are reported in xenophyophores, a group of rhizopod protozoans found at abyssal and bathyal depths in the oceans (35 000 ppm ashed carbonate-free residue) (Gooday & Nott, 1982). Further, barite crystals and granules (in the size range 1–5 μm) occur within the cells of certain algae (Kreger & Boere, 1969; Schröter *et al.*, 1975; Fresnel *et al.*, 1979), in certain ciliates (Hubert *et al.*, 1975) and in xenophyopores (Protozoa) (Gooday & Nott, 1982). However, such biogenically precipitated barite is unlikely to make a direct contribution to the barite in marine sediments as the solubility of barium sulphate is much higher in the presence of organic matter than in its absence (Desai *et al.*, 1969) and most such biogenic barite probably dissolves on contact with seawater. However, barium extracted from seawater by marine organisms is released at depth on dissolution of their remains (Church, 1970). This would lead to enrichment of barium in bottom waters and the interstitial waters of underlying sediments. Progressive enrichment could lead to barium saturation and its precipitation (Church, 1970). Such processes could explain the high concentrations of barite found in sediments beneath the equatorial high productivity zone (Arrhenius & Bonatti, 1965; Church, 1970; Chesselet *et al.*, 1976). Chesselet *et al.* (1976) report the occurrence of particulate barite in ocean waters. This is probably authigenic in origin and on settling may contribute to barite concentration in sediments.

There is mounting evidence that barite formation is not restricted to the water column or subsea sediment surface but can continue within sediments after their burial (Church & Bernat, 1972; Cronan, 1974; Dean & Schreiber, 1977).

Main Distinguishing Features

Form/shape, cleavage, weak birefringence and straight extinction. Barite can resemble anhydrite but has a lower birefringence and different cleavage.

CALCITE

Biogenic/detrital/authigenic
$CaCO_3$
Calcium carbonate

Form/Shape

Biogenic calcite occurs as chambered foraminifera tests and test fragments (Figs 13–15), ovoid coccolith and stellar discoaster plates (Figs 16–19). Foraminifera chamber forms and arrangements are extremely varied (see Appendix 5). Most of the formanifera tests accumulating on the deep seafloor are of planktonic species which, in life, inhabit mainly the upper 200–300 m of the water column. Most planktonic foraminifera have multichambered tests, the chambers being spirally and/or cyclically arranged. Test walls are frequently perforated by numerous pores. Coccoliths are minute oval plates of calcite produced by unicellular planktonic algae. Most are formed of minute rhombohedral crystals, although a few have hexagonal prisms. Discoasters are tiny star or rosette-shaped plates. They have tabular crystalline forms and are more coarsely constructed than coccoliths. High magnifications (×1000) will be needed to resolve individual coccolith and discoaster plates. Non-skeletal calcite debris may occur as lumps, grains, ooids, aggregates and pellets (Fig. 20).

Grain Size Range

Very variable, most coccolith and discoaster plates fall into the range 2 to 35 μm and most foraminifera are in the range 100 to 200 μm. Discoasters and coccoliths are collectively termed nannofossils as they are rather near the limit of resolution of light microscopy and hence best studied using the electron microscope. Detrital carbonate debris may vary from fine silt to very coarse sand (4–1000 μm).

Transparency/Colour in PPL

Generally colourless and transparent. Biogenic calcite may show a granular texture and show pores or other kinds of surface sculpture.

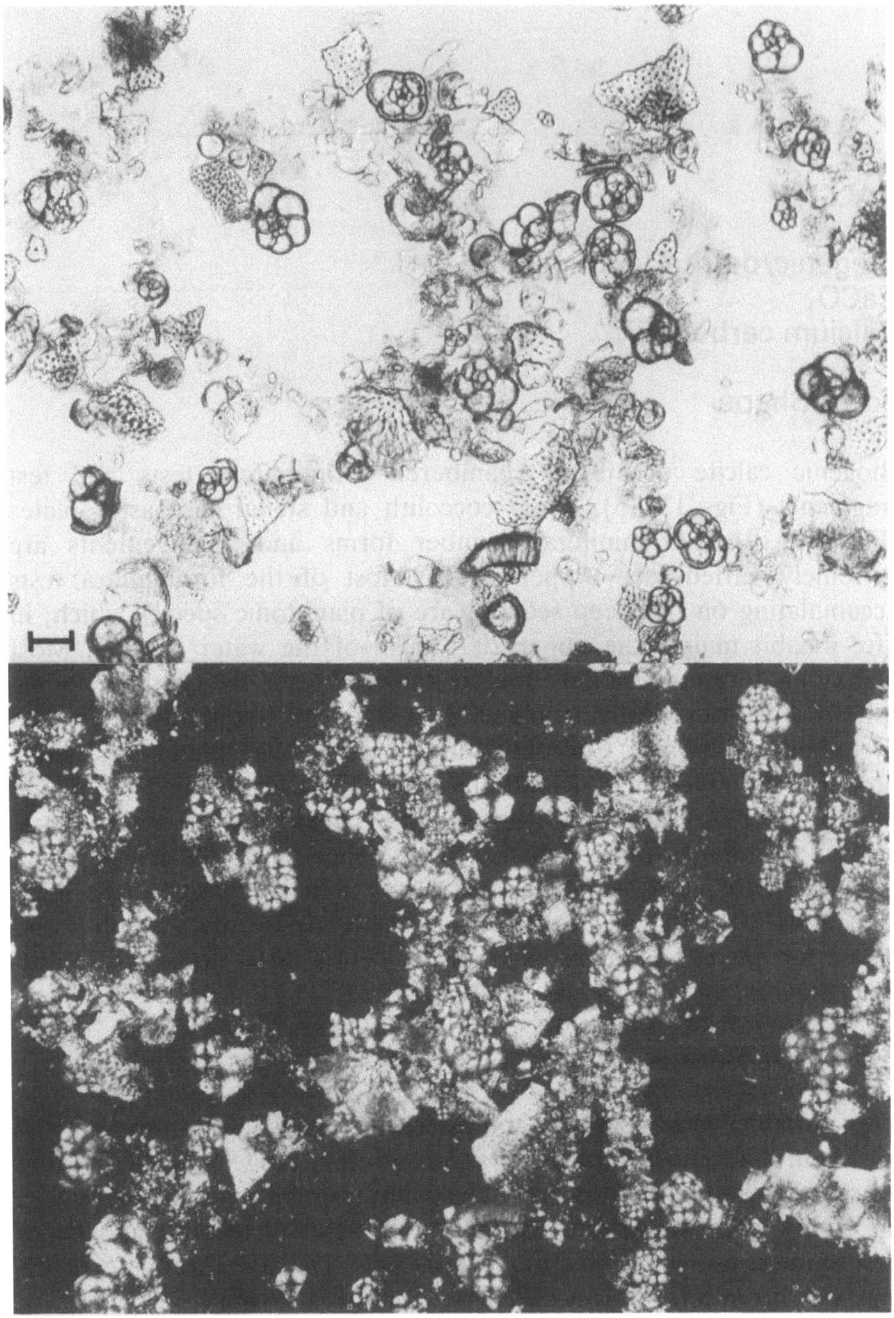

Fig. 13. Biogenic calcite largely consisting of foraminifera tests and test fragments seen in plane-polarised light (top) and in cross-polarised light (bottom). Note the black interference crosses seen on each test chamber in the cross-polarised view. Pleistocene, Kings Trough, Northeast Atlantic (43°52′N, 21°57′W). Scale bars equal 0·05 mm.

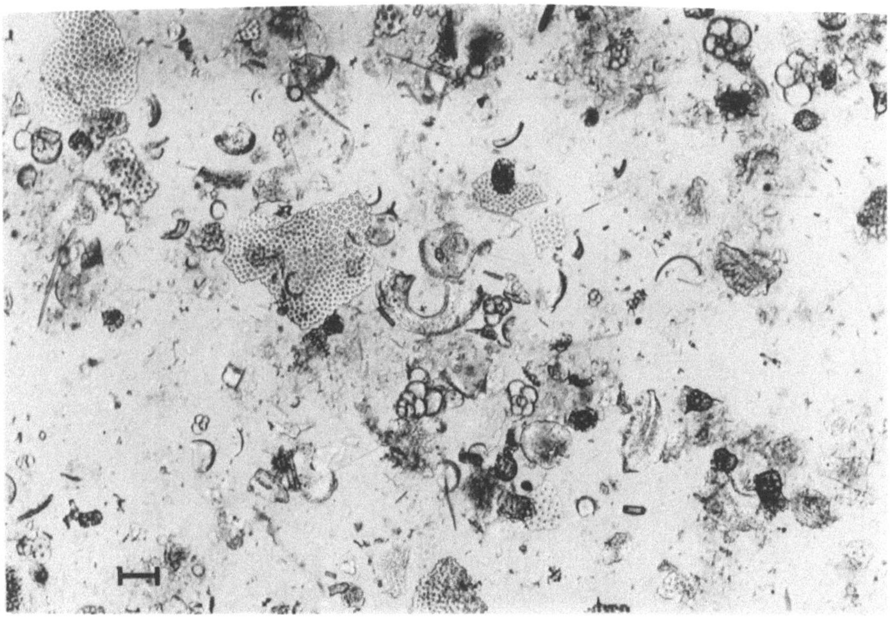

Fig. 14. Calcareous ooze comprising largely of foraminifera tests and test fragments (biogenic calcite). Pleistocene, Amirantes Passage, Indian Ocean (9°28′S, 52°04′E). Plane-polarised light, scale bar equals 0·05 mm.

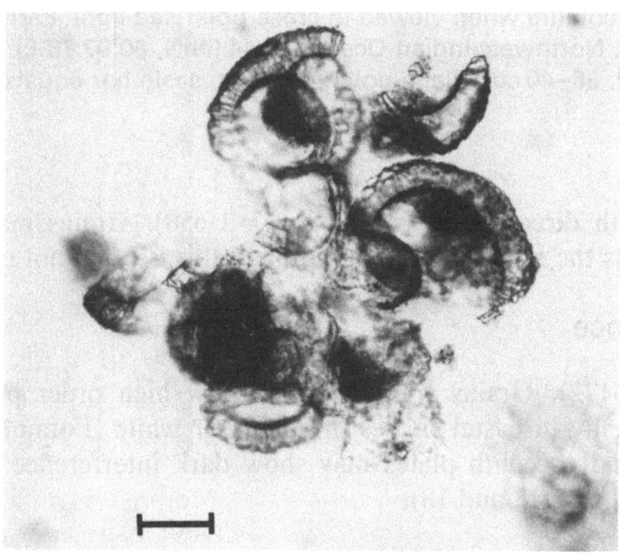

Fig. 15. Crushed foraminifer. Pleistocene, Kings Trough, Northeast Atlantic (43°50′N, 21°51′W). Plane-polarised light, scale equals 0·05 mm.

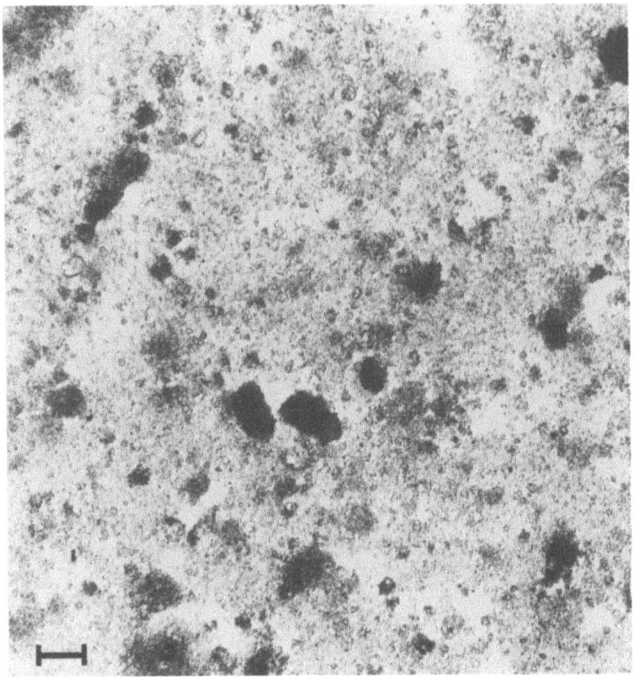

Fig. 16. Fine-grained calcite largely consisting of coccolith plates. Such material can easily be distinguished from clay by its bright high order interference colours when viewed in cross-polarised light. Early Miocene, Owen Ridge, Northwest Indian Ocean (18°44.98′N, 60°07.78′E). DSDP Leg 23, 223-27-2, 38–40 cm. Plane-polarised light, scale bar equals 0·05 mm.

Relief

Variable with direction ($n_\varepsilon = 1{\cdot}486$, $n_\omega = 1{\cdot}658$). Grains may show a 'twinkling' as the stage is rotated, although this is often not obvious.

Birefringence

Extreme (0·172). Grains will therefore show high order polarisation colours, usually in pastel shades of yellow or white. Foraminifera test chambers and coccolith plates may show dark interference crosses in cross-polars (Figs 13 and 18).

Alteration

Calcite is stable on the ocean floor.

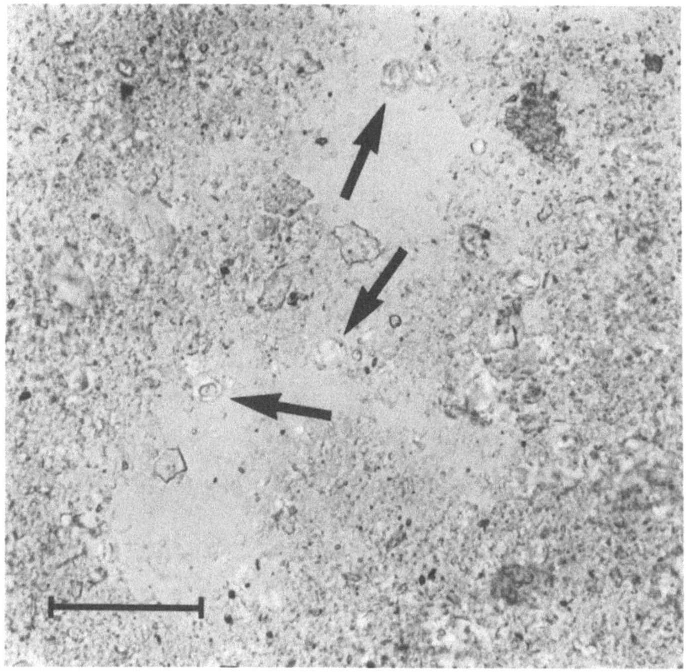

Fig. 17. Fine-grained calcite seen under high magnification. The material is largely composed of entire and broken coccolith plates. Three entire plates are arrowed. Pleistocene, Kings Trough, Northeast Atlantic (41°28.5′N, 21°41.4′W). Plane-polarised light, scale bar equals 0·05 mm.

Crystal System

Trigonal.

Cleavage

$\{10\bar{1}1\}$ perfect. There is sometimes parting along twin lamellae on $\{0\bar{1}12\}$.

Pleochroism

Absent.

Extinction Angle

In calcite crystals, extinction is symmetrical to the cleavage traces. Biogenic debris will not extinguish as the stage is rotated, although the

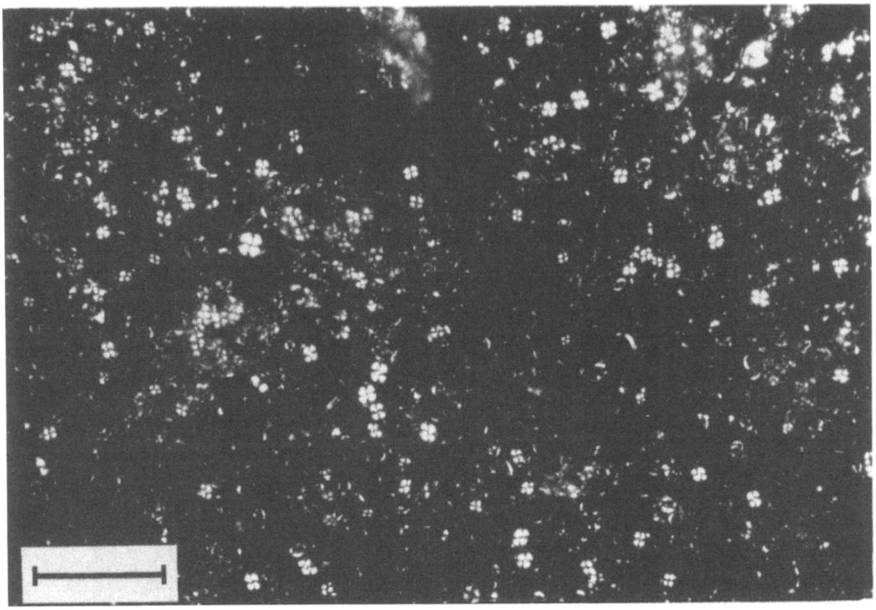

Fig. 18. Scatter of coccolith plates seen in cross-polarised light. Note the black interference crosses shown by each plate. Early Cretaceous (Aptian), lower continental rise southeast of New York, Northwest Atlantic (34°54′N, 69°10′W). DSDP Leg 11, 105-13-CC (core-catcher). Scale bars equals 0·05 mm.

dark interference crosses will appear to rotate, but actually remain stationary with respect to the direction of polarisation.

Twinning

Non-skeletal calcite grains may show polysynthetic twinning with $\{01\bar{1}2\}$ as the twin plane.

Optical Character

Calcite has a uniaxial negative interference figure, although the figure may be very eccentric.

Occurrence

Calcite shows increasing solubility with depth and this is related to increased hydrostatic pressure, increasing CO_2 content within the

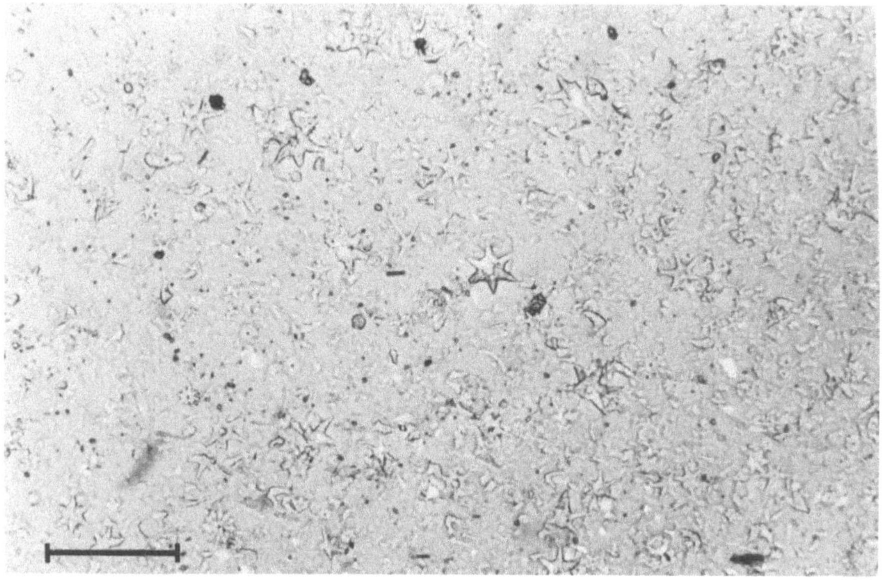

Fig. 19. Field of discoasters—stellar calcareous plates produced by a now extinct group of Tertiary nannoplankton. Middle Eocene, Cantabria Seamount, Bay of Biscay, Northeast Atlantic (45°01.9′N, 7°58.49′W). DSDP Leg 12, 119-22-2, 9 cm. Plane-polarised light, scale bar 0·05 mm.

ambient water and decreasing temperature (Milliman, 1974). Consequently, a depth level is present within the water column which separates well-preserved from poorly-preserved, solution-etched foraminifera and coccoliths. Berger (1968) has termed this depth level, the lysocline. The position of the lysocline varies but generally lies at water depths of between 3000 and 5500 m (Berger, 1976). At greater depths, the rate of supply of biogenic carbonate equals its rate of dissolution. Bramlette (1961) called this depth the calcite compensation depth (CCD). The regional depth of the CCD is a function of a number of variables:

—the rate of supply of biogenic carbonate from surface waters;
—the depth of the lysocline; and
—the dissolution rate gradient below the lysocline.

These factors will reflect ocean productivity patterns and the shoaling of the lysocline near continental margins (Andel *et al.*, 1975). The CCD varies from between 3500 and 5500 m water depth in the Atlantic and

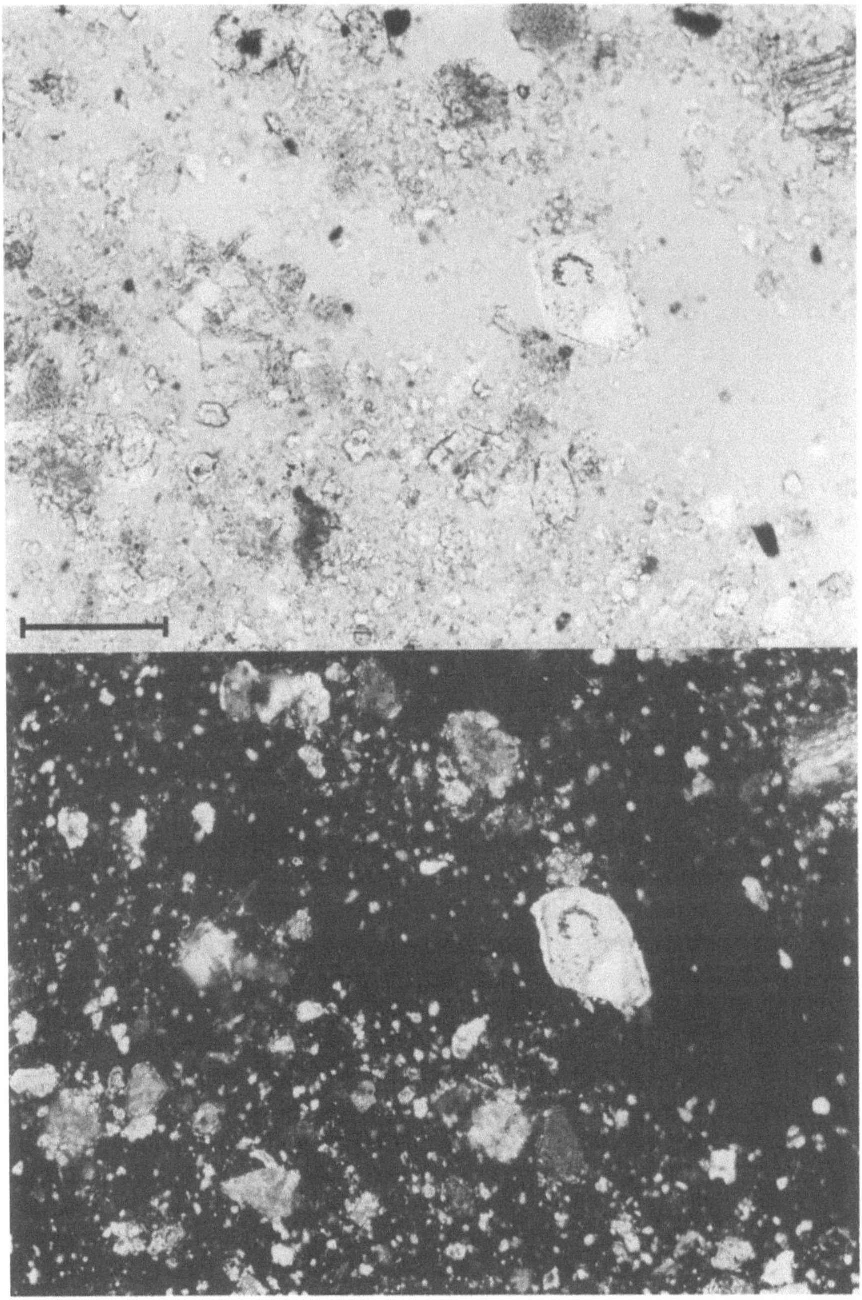

Fig. 20. Granular calcite, viewed in plane-polarised light (top) and cross-polarised light (bottom). Pleistocene, Tyrrhenian Sea, Mediterranean Sea. Scale bar equals 0·05 mm.

between 3000 and 5000 m in the Pacific (Berger, 1976). Thus the occurrence of calcite in deep-sea sediments will be depth-controlled. During the Cenozoic and Mesozoic, the CCD has shown global fluctuation (Andel, 1975). During the Cenozoic, this reflected glacial-related changes in bottom water structure but prior to about 50 Ma, the CCD depth probably reflected large-scale changes in the distribution of carbonate deposition between shallow and deep areas, or a low input of calcium to the oceans from the continents (Andel, 1975).

As deep ocean water is undersaturated with respect to its low temperature and high CO_2 concentration, direct precipitation of carbonates is rare in the deep sea. However, Bonatti (1966) reports authigenic calcite and dolomite crystals up to 90 μm in size from red clay in the South Pacific, probably formed as a result of hydrothermal activity in the seafloor.

Oceanic Distribution

Mid-ocean ridge flanks, aseismic ridges and continental rises. Its distribution pattern generally reflects the depth of the CCD in relation to bathymetry and terrigenous sediment input. In the Atlantic Ocean, calcareous oozes occur principally along the flanks of the Mid-Atlantic Ridge, the Walvis Ridge, the Rio Grande Rise and on the continental rise. In the Pacific Ocean, calcareous oozes occur principally on the flanks of the East Pacific Rise, and in the Western South Pacific. In the Indian Ocean, calcareous oozes are largely restricted to the Mid-Indian Ocean and Southwest Indian Ridges, the Ninetyeast Ridge, the Madagascar Plateau and the Naturaliste and Wallaby plateauxs off eastern Australia.

Source/Transport Mechanisms

Calcite will be supplied to the deep ocean floor through the constant 'rain' of coccolith and foraminifera tests from the upper waters of the oceans, following death of the organism. Non-skeletal carbonate debris, derived from shallow water and the erosion of carbonate sediments, will be transported to the deep ocean mainly by mass gravity flows, ice-rafting and aeolian deposition.

Diagenesis

During diagenesis, calcite grains will become progressively covered by subhedral to euhedral calcite cement. The perforations of coccolith plates, the central areas of coccospheres and foraminifera chambers become filled with granular calcite and large amounts of interparticle cement are formed. Grain interpenetration and welding becomes common. Continued diagenesis will result in a 'pavement mosaic' texture of completely interlocking grains (Fischer *et al.*, 1967).

Main Distinguishing Features

Form, occurrence as skeletal debris, extreme birefringence, common occurrence.

Appearance as Loose Grains

Depending on their origin, calcite allochems will show some morphological variety: occurring as globular single or multichambered foraminifera tests; spherical ooids of 0·1–1·0 mm diameter; ovoid rounded faecal pellets, averaging 0·03–0·20 mm in length; irregular non-skeletal debris of varied size and shape and fine-grained 'amorphous' calcite. Amorphous calcite, representing clumps of coccolith debris, will comprise the main form when calcareous oozes are viewed in reflected light. Calcite will appear white with a vitreous, porcellanous or dull lustre depending on its form.

Appearance in Thin Section

Plane-polarised light: colourless but often cloudy, fine to coarse anhedral aggregates or organic debris. Grains often show an organic structure. Variable relief. Cross-polarised light: extreme birefringence, the maximum interference colours being high order whites. Extinction is symmetrical to the cleavage traces.

CLAY

Detrital and authigenic

Clay is a textural term referring to particulate material under 4 μm in size (Wentworth, 1922). A proportion of this material will be clay minerals, an important group of silicate layer–lattice minerals which occur as minute platey, or more rarely, fibrous crystals. Their very fine grain size precludes any determinative work using smear slides.

Form/Shape

Clay will appear as unsorted irregular aggregates or clumps of numerous individual grains which may or may not incorporate organic debris (Figs 21 and 22). The clumps are very variable in size and shape and most consist of small transparent–translucent mineral grains.

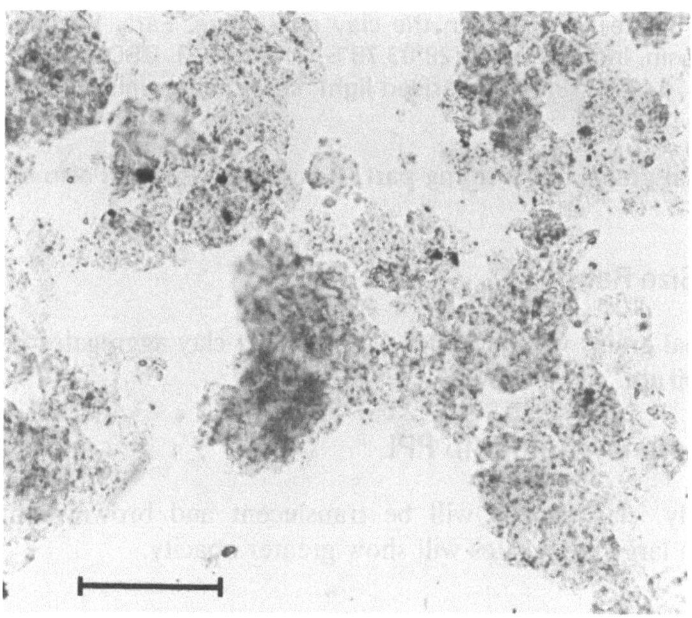

Fig. 21. Irregular aggregates and clumps of clay grade material. Pleistocene, Madeira Abyssal Plain, Northeast Atlantic. Plane-polarised light, scale bar equals 0·05 mm.

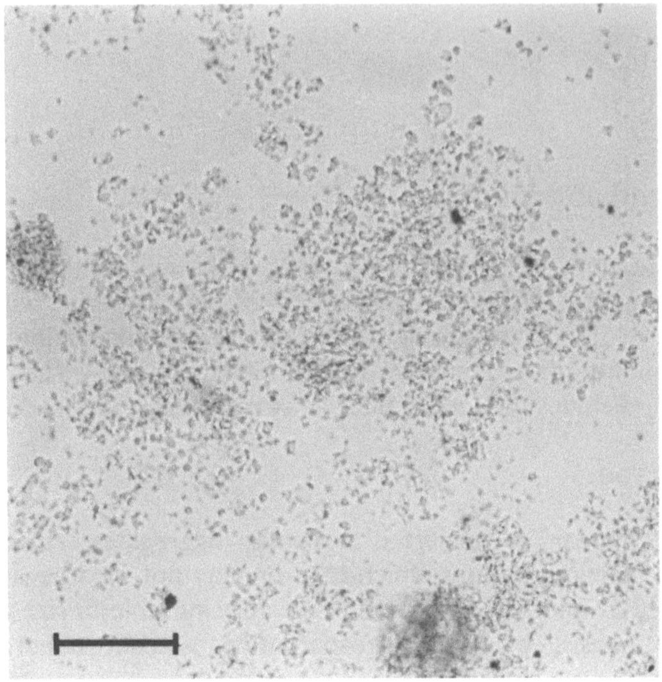

Fig. 22. Scatter of grains in the clay size range. Early Eocene, Mozambique Basin, Indian Ocean (29°03.78′S, 37°28.48′E). DSDP Leg 25, 248-11-1, 140 cm. Plane-polarised light, scale bar equals 0·05 mm.

Individual grains not forming part of larger masses will also occur (Fig. 22).

Grain Size Range

Individual grains will be under 4 μm in size, clay aggregates may range up to 200 μm.

Transparency/Colour in PPL

Generally, clay clumps will be translucent and brownish in colour, although large aggregates will show greater opacity.

Relief

Cannot be quantified. However, the relative size and thickness of the clay masses may give a high apparent relief.

Birefringence

Clay clumps will show aggregate polarisation with low first order–low second order interference colours which will usually be masked by body colour.

Occurrence

Clay is ubiquitous throughout the marine realm. Clay is derived through continental weathering and by diagenetic alteration of a wide range of other sedimentary materials (e.g. volcanic glass, palagonite, feldspars). X-ray diffraction studies have shown that the bulk of Recent deep-sea clay is derived from the continents (Biscaye, 1965). Biscaye (1965) has also shown that mineralogical anlayses of the fine fraction of deep-sea sediments are a useful indicator of sediment provenance.

Main Distinguishing Features

Form, fine grain size, aggregate polarisation in cross-polarised light.

Notes/Remarks

Clay minerals in marine sediments can be useful indicators of palaeoclimate (Chamley, 1967). Although they do not provide direct indications of climatic parameters, clay minerals can provide integrated records of overall climatic impact which may occasionally be superior to more conventional palaeoclimatic indicators such as pollen or oxygen isotope analysis. Such analysis relies on the identification of specific clay minerals using X-ray diffraction, so it lies outside use with smear slides. A recent review is provided by Singer (1984).

DOLOMITE

Authigenic
$CaMg(CO_3)_2$
Calcium magnesium carbonate

Form/Shape

Generally occurs as small rhombohedral grains (Figs 23–26), but may more rarely occur as tiny waxy spheroids aggregated into cylindrical masses, suggestive of worm tubes. These grains are possibly otoliths (ear-bones of fish) in which dolomite has replaced the original calcite.

Grain Size Range

Generally small, being in the fine-medium silt range (15 μm or smaller).

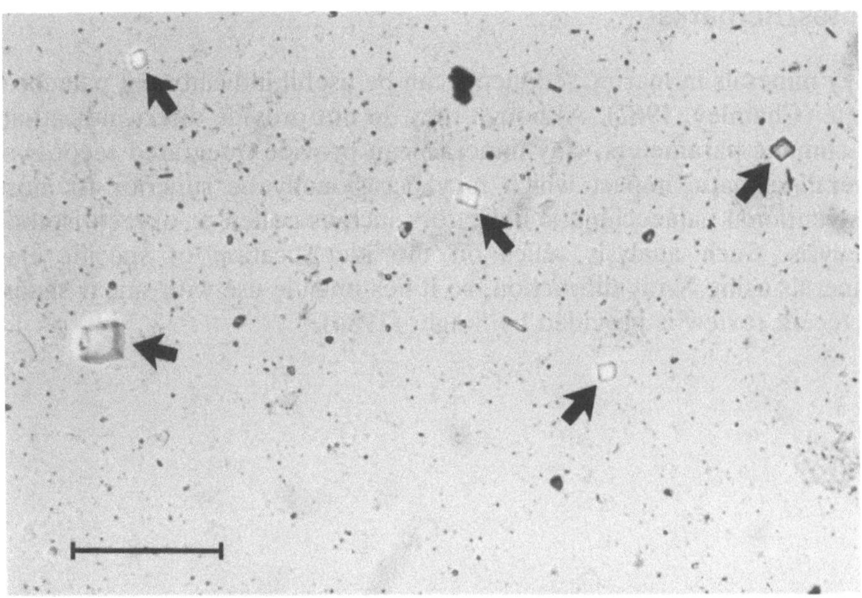

Fig. 23. Scatter of dolomite rhombs (arrowed). Early Eocene, Cantabria Seamount, Bay of Biscay, Northeast Atlantic (45°01.9′N, 7°58.49′W). DSDP Leg 12, 119-24-3, 95 cm. Plane-polarised light, scale bar equals 0·05 mm.

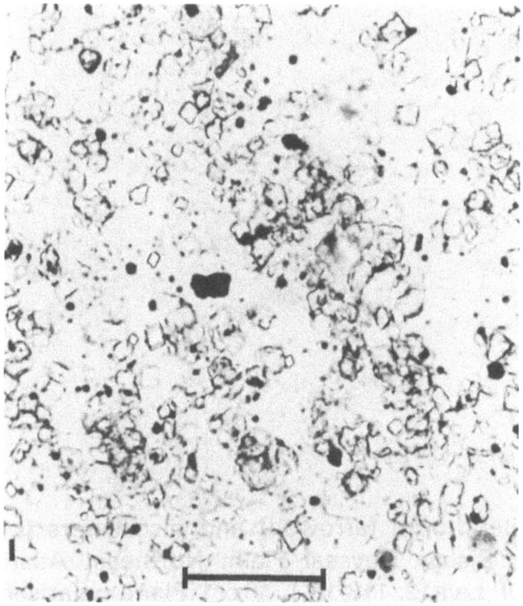

Fig. 24. Rhombohedral grains of dolomite. Small amounts of pyrite are also present. Pleistocene, Cariaco Trench, Venezuelan continental shelf (10°42.48′N, 65°10.48′W). DSDP Leg 15, 147-9-CC (core-catcher). Plane-polarised light, scale bar equals 0·05 mm. (Courtesy of Dr T. W. Donnelly and the Deep Sea Drilling Project.)

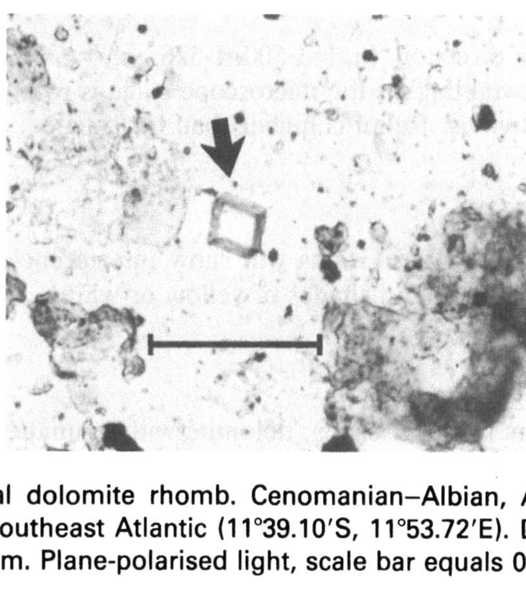

Fig. 25. Euhedral dolomite rhomb. Cenomanian–Albian, Angolan continental slope, Southeast Atlantic (11°39.10′S, 11°53.72′E). DSDP Leg 40, 365-2-3, 8 cm. Plane-polarised light, scale bar equals 0·05 mm.

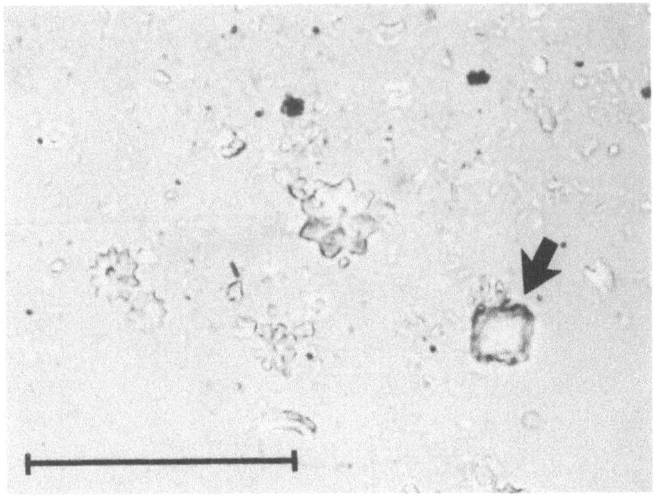

Fig. 26. Dolomite rhomb (arrowed) and discoasters (biogenic calcite). Middle Eocene, Biscay Abyssal Plain, Northeast Atlantic (45°02.65′N, 9°00.63′W). DSDP Leg 12, 118-13-1, 46 cm. Plane-polarised light, scale bar equals 0·05 mm.

Transparency/Colour in PPL

Generally colourless and transparent.

Relief

Variable with direction ($n_\varepsilon = 1{\cdot}500$–$1{\cdot}526$, $n_\omega = 1{\cdot}680$–$1{\cdot}716$). Grains may show a 'twinkling' as the microscope stage is rotated, although this may not be obvious. Relief is higher than for calcite.

Birefringence

Extreme (0·180–0·190). Grains will show interference colours of high order, most frequently in shades of yellow or white.

Alteration

When it occurs in smear slides, dolomite will be unaltered.

Crystal System

Trigonal.

Cleavage

Perfect rhombohedral parallel to $\{10\bar{1}1\}$, most often showing as two sets of lines intersecting at an oblique angle.

Pleochroism

Absent.

Extinction Angle

Symmetrical to the crystal outlines and to the cleavage traces.

Optical Character

Dolomite has a uniaxial negative interference figure which may show many rings.

Associated Minerals

Often associated with pyrite, reflecting the specific environmental conditions required for its formation (see below).

Occurrence

Dolomite is relatively uncommon in modern ocean sediments, even though thermodynamic calculations suggest that dolomite should be the most stable carbonate mineral in seawater. The genesis of dolomite in ocean sediments is still problematic. Dolomite is not known to be precipitating from seawater of normal salinity ($35^{oo}/_{oo}$) anywhere in the world at present. Although the direct precipitation of dolomite from seawater has yet to be demonstrated, it has been postulated for several dolomites in the geological record, including the Mediterranean Messinian dolomites cored during Leg 13 of the Deep Sea Drilling Project (Ryan *et al.*, 1973); enabled perhaps by elevated salinities in shallow water (e.g. Behrens & Land, 1972; Nesteroff, 1973). Recent research indicates that the genesis of dolomite is not controlled primarily by the Mg^{2+}/Ca^{2+} ratio in seawater as previously thought but by dissolved sulphate which inhibits the formation of dolomite. In its absence, in experimental studies seawater reacts rapidly with both calcite and aragonite to form dolomite. In the ocean basins, the most common process for such removal of sulphate from seawater is its bacterial reduction to hydrogen sulphide within marine sediments rich in organic matter. Such environments occur in many near shore basins such as the Gulf of California, where dolomitic sediments have recently been described (Odegard, 1981). In the highly organic Pleistocene sediments of the Cariaco Trench in the southwestern Columbian basin (DSDP Hole 147), almost monomineralic beds of dolomite occur (Donnelly & Nalli, 1973), presumably formed through the removal of inhibitory sulphate due to bacterial reduction. A spectacular recovery of dolomite occurred during Ocean Drilling Program Leg 112, when dolomite was found to be almost ubiquitous in organic-rich, diatomaceous sediments on the Peru margin and dolomite nodules were found within a few metres of the seabed. Excellent geochemical data was also obtained supporting sulphate control of dolomite formation—dolomite rhombs were found in sediments where SO_4^{2-} values in the interstitial water reached zero (Hill, P., 1987 pers. comm.).

In marine sediments, dolomite is nearly always authigenic in origin. Derived dolomite tends to be rare but may occur in some locally derived terrigenous sediments. Such dolomite usually has a source in local dolomitic limestones.

Main Distinguishing Features

Characteristic rhombohedral form, 'twinkling' as the microscope's stage is rotated and typically small size range are the best criteria. However, dolomite may sometimes be difficult to distinguish on account of its small grain size and similarity to calcite rhombs. In theory, dolomite rhombohedra should be easily distinguishable from calcite rhombohedra by their higher refractive index. However, calcite rhombs may occur as pseudo-twinned rhomb aggregates which, in smear slides, tend to lie with their *c* axis vertical, showing a very high refractive index in all directions (Donnelly & Nalli, 1973).

FELDSPAR

Detrital and authigenic

Although closely related, the feldspars fall into two sub-groups: (1) the alkali feldspars which are generally monoclinic and (2) the sodium and calcium feldspars which form a solid solution series, collectively known as the plagioclase feldspars, which are triclinic. Members of the plagioclase series thus range in composition from 100% albite ($NaAlSi_3O_8$), often abbreviated Ab, to 100% anorthite ($CaAl_2Si_2O_8$), abbreviated An. Within the plagioclase series, a number of feldspar types are distinguished based on their percentage composition of albite/anorthite (Table 6).

TABLE 6 The Feldspar Group

THE ALKALI FELDSPARS			
Monoclinic			
Orthoclase			$(KNa)AlSi_3O_8$
Sanidine			$(KNa)AlSi_3O_8$
Adularia			$(KNa)AlSi_3O_8$
Triclinic			
Microcline			$(KNa)AlSi_3O_8$
Anorthoclase			$(NaK)AlSi_3O_8$
THE PLAGIOCLASE SERIES			
Triclinic			
Albite	$NaAlSi_3O_8$	Composition	90–100% Ab, 0–10% An
Oligoclase			70–90% Ab, 10–30% An
Andesine			50–70% Ab, 30–50% An
Labradorite			30–50% Ab, 50–70% An
Bytownite			10–30% Ab, 70–90% An
Anorthite	$CaAl_2Si_2O_8$		0–10% Ab, 90–100% An

Form/Shape

Subhedral prismatic or tabular to rounded anhedral grains or cleavage fragments (Figs 27–30).

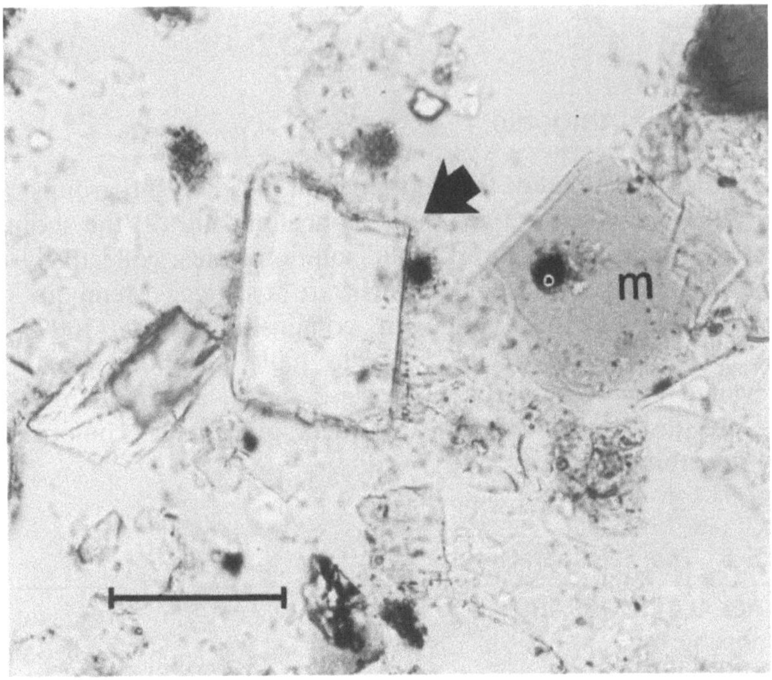

Fig. 27. Subhedral alkali feldspar crystal (arrowed). A mica flake (m) can be seen to the right. Pleistocene, Tyrrhenian Sea, Mediterranean. Plane-polarised light, scale bar equals 0·05 mm.

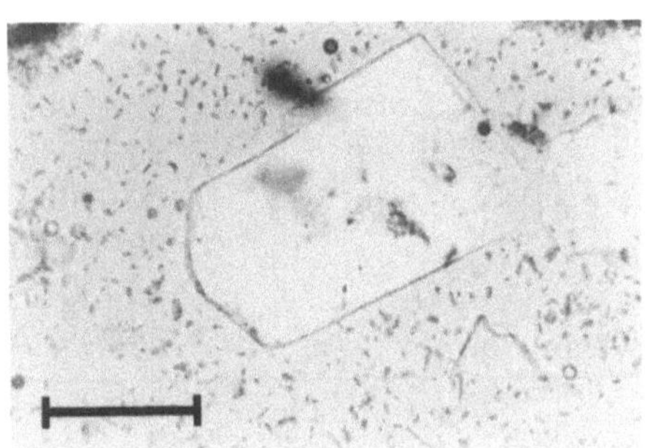

Fig. 28. Subhedral crystal of alkali feldspar. Pleistocene, Tyrrhenian Sea, Mediterranean. Plane-polarised light, scale bar equals 0·05 mm.

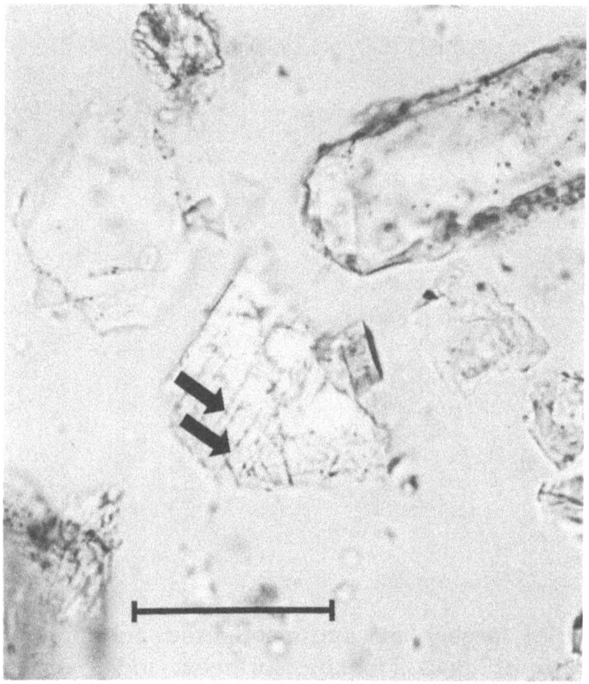

Fig. 29. Subhedral fragment of plagioclase feldspar. The lamellar twin planes (arrows) are just visible. Pleistocene, Tyrrhenian Sea, Mediterranean. Plane-polarised light, scale bar equals 0·05 mm.

Fragmentation Characteristics

Cleavage fragments will appear as small flat plates having two sides parallel.

Grain Size Range

2–500 μm. However, it is likely that only grains in the medium-silt to sand-size range will be recognisable in smear slides.

Transparency/Colour in PPL

Colourless but frequently cloudy due to incipient alteration (this characteristic can be used to distinguish from quartz) (Fig. 31). The variety sanidine is nearly always clear and shows little alteration. Sanidine, however, may be distinguished from quartz in smear slides by its crystal form, usually being subhedral, and its perfect cleavage.

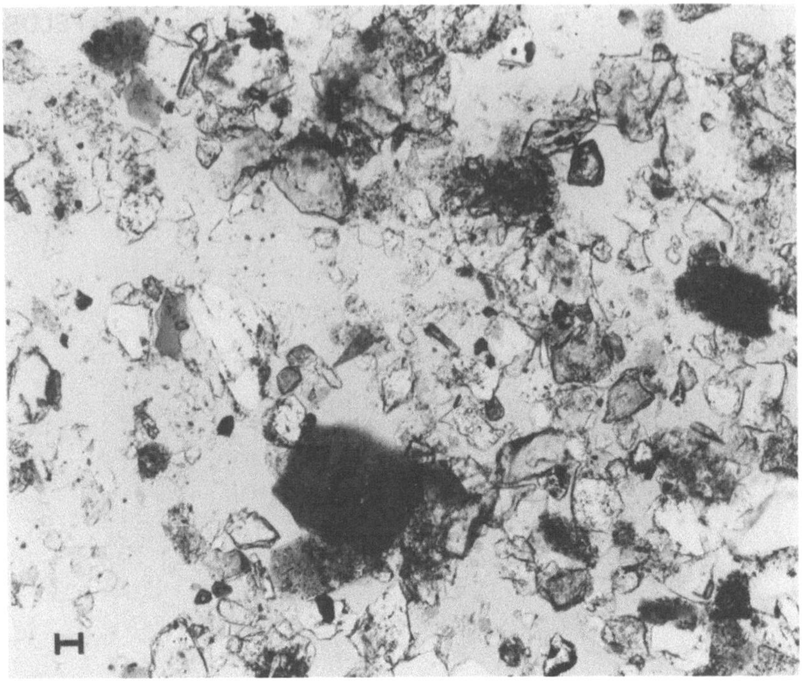

Fig. 30. Field of largely silt and sand-sized anhedral alkali feldspar. Pliocene, south of Java Trench, Northeast Indian Ocean (9°46.53′S, 102°41.95′E). DSDP Leg 22, 211-8-1, 110 cm. Plane-polarised light, scale bar equals 0·05 mm.

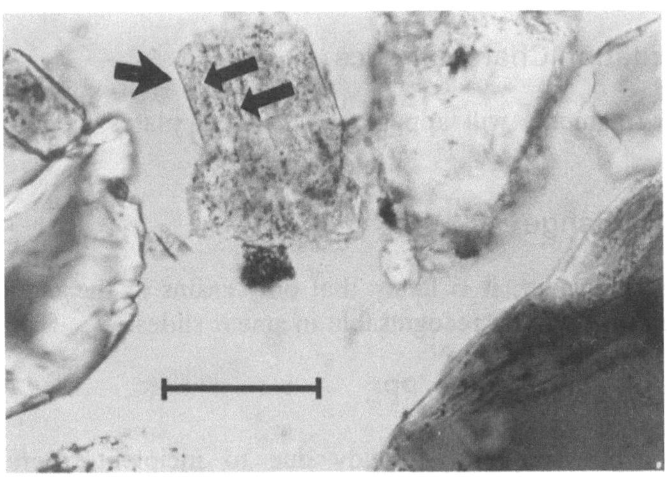

Fig. 31. Subhedral plagioclase crystal (arrowed) showing traces of lamellar twinning (small arrows). The crystal appears turbid due to incipient alteration. Pleistocene, Tyrrhenian Sea, Mediterranean. Plane-polarised light, scale bar equals 0·05 mm.

Relief

Relief varies with composition. The refractive indices are as follows:

The alkali feldspars

	n_α	n_β	n_γ
Orthoclase	1·518	1·524	1·526
Sanidine	1·517–1·520	1·523–1·525	1·524–1·526
Microcline	1·518–1·522	1·522–1·526	1·525–1·530
Anorthoclase	1·522–1·536	1·526–1·539	1·527–1·541

The plagioclase feldspars

	n_α	n_β	n_γ
Albite	1·527–1·533	1·531–1·537	1·538–1·542
Oligoclase	1·533–1·543	1·537–1·548	1·542–1·551
Andesine	1·543–1·554	1·548–1·558	1·551–1·562
Labradorite	1·554–1·564	1·558–1·569	1·562–1·573
Bytownite	1·564–1·573	1·569–1·579	1·573–1·585
Anorthite	1·573–1·577	1·579–1·585	1·585–1·590

All the alkali feldspars and plagioclase feldspars more sodic than An_{30} have refractive indices lower than quartz, while plagioclases more calcic than An_{30} have indices higher than quartz. Thus relief of most feldspars is low.

Birefringence

All feldspars show weak birefringence (0·005–0·011) which in all the common feldspars is lower than quartz, so that interference colours shown by silt-sized grains will be low order. Crystals may be compositionally zoned (Fig. 32).

Alteration

Feldspars are unstable under most conditions and alter to clay minerals. So feldspar grains seen in smear slides may be turbid due to alteration and this may sometimes obscure other optical properties (such as visibility of twinning). Microcline and sanidine are the most stable and

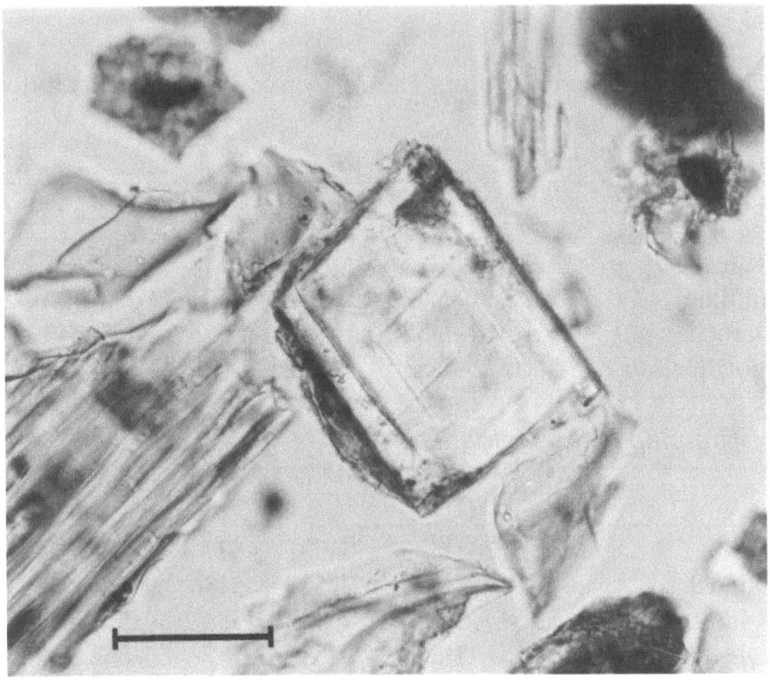

Fig. 32. Euhedral feldspar crystal showing oscillatory zoning. In cross-polarised light, such oscillatory zoned crystals will display concentric shells of varying extinction. Oscillatory zoning usually reflects varying water vapour pressures in the original melt, frequently due to the periodic escape of gases. Such crystals usually have a volcanic origin. The thin envelope of glass around this particular crystal shows it is derived from pyroclastics. Pleistocene, Tyrrhenian Sea, Mediterranean. Plane-polarised light, scale bar equals 0·05 mm.

sanidines will appear clear and usually free of bubble inclusions. Orthoclase is the next most stable, while sodic plagioclase is more unstable and calcic plagioclase is very unstable (Goldich, 1938) (Table 5).

Cleavage

Both the alkali and plagioclase feldspars have good cleavage in two directions, parallel to {001} and {010}.

Pleochroism

Absent.

Extinction angle

In the alkali feldspars, extinction is parallel on {001} and up to 12° on {010}. In the plagioclases, the extinction angle varies with composition and measurement of this forms the basis of the Michel–Levy method for discriminating plagioclase feldspars.

The Michel–Levy method relies on the measurement of the maximum extinction angle in the zone of faces perpendicular to (010). Thus it is necessary to identify grains which lie with the (010) plane normal to the plane of the slide. Such grains are identifiable as they will show clear, sharp polysynthetic twin lamellae of the albite type (Fig. 33).

Michel–Levy method for discriminating plagioclase feldspars:

1. View slide in cross-polarised light. If a crystal shows clear lamellar twinning of the albite type, then it lies with the {010} crystal plane normal to the plane of the slide and is suitable for use.

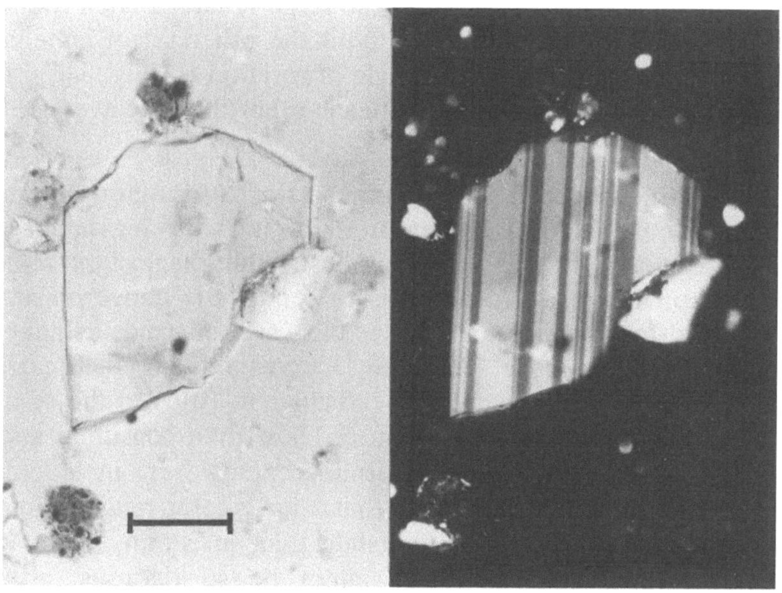

Fig. 33. Subhedral plagioclase crystal showing polysynthetic twinning on the albite law. The crystal is viewed in plane-polarised light (left) and cross-polarised light (right). Traces of the lamellar twinning, so obvious in cross-polars, can just be discerned in the plane-polarised view. Pliocene, south of Java Trench, Northeast Indian Ocean (9°46.53′S, 102°41.95′E). DSDP Leg 22, 211-8-1, 56 cm. Scale bar equals 0·05 mm.

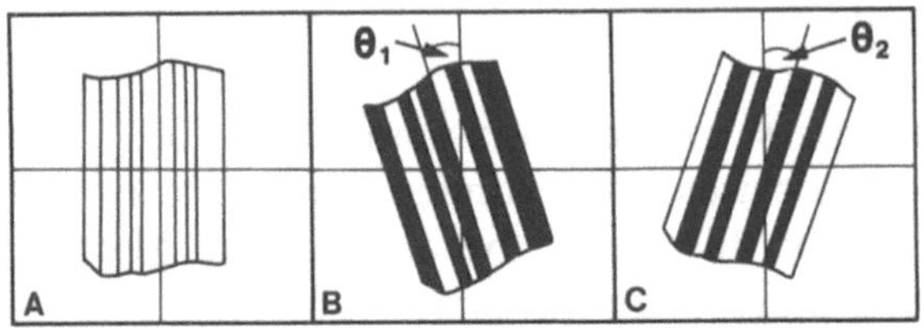

Fig. 34. Michel–Levy method for discriminating plagioclase feldspars. See text for method.

2. Rotate the stage so that all the lamellae are equally illuminated when parallel to one of the crosswires. Note reading on the stage to nearest degree (Fig. 34A).
3. Rotate stage clockwise or anticlockwise until one set of lamellae is at extinction and note new reading on stage. Subtract this new reading from that noted when lamellae were equally illuminated to find the angle of rotation θ_1 (see Fig. 34B).
4. Rotate stage back to the position where all the lamellae are equally illuminated, again note the stage reading. Then rotate the stage in the opposite direction until the other set of lamellae go into extinction. Measure the angle of rotation θ_2 (Fig. 34C). If the angles θ_1 and θ_2 are within 5° of each other then the average value is taken.

When working with solid rocks in thin section, it is important that the maximum extinction angle is measured for it is the maximum value which is diagnostic of the composition of the plagioclase feldspar. Therefore, conventional procedure is to repeat the above on at least five different feldspar crystals in a thin section and to take the maximum extinction angle measured. Curves relating the maximum extinction angle in sections cut perpendicular to (010) to the composition of the plagioclase feldspar (Fig. 35) are then consulted and the composition of the plagioclase determined. However, in a feldspar-bearing marine sediment, it is likely that a far greater range of feldspar composition may be present in the slide than in a thin section of an igneous rock. Consequently, one cannot be so rigorous, although generally, the Michel–Levy method will give a broad indication of the composition of any one particular plagioclase feldspar. Reference to Fig. 35 shows albite, oligoclase and andesine all have maximum extinction angles of less than 20°. However, plagioclases with a composition between An_0 to An_{20} (albite, oligoclase) have refractive indices lower than Canada balsam (and quartz) while plagioclases

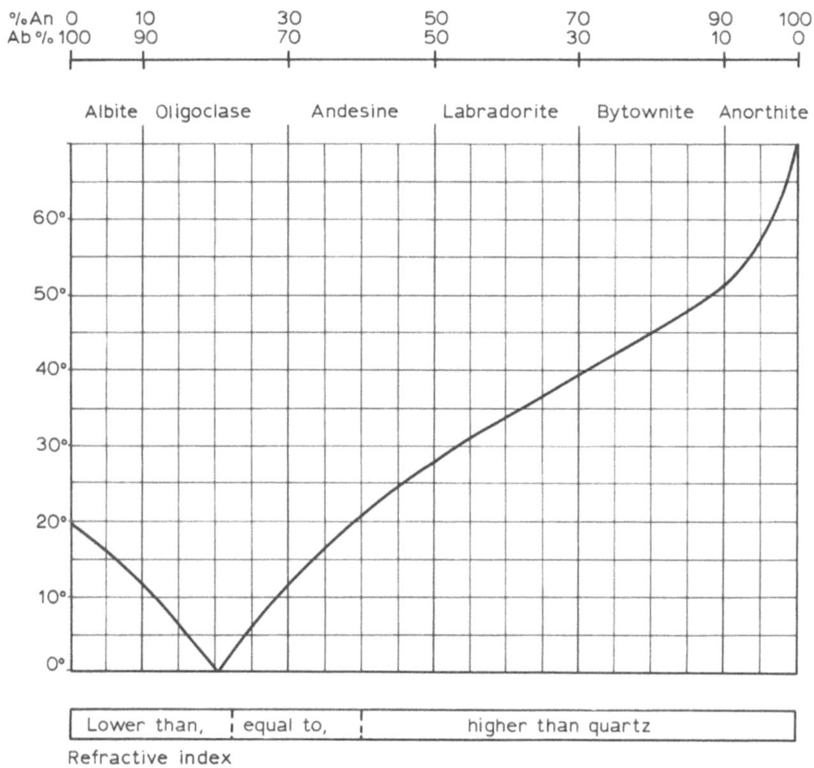

Fig. 35. Curves showing the maximum extinction angle of albite twin lamellae of plagioclase cut normal to {010} (after Kerr, 1977).

having a composition of between An_{20} to An_{100} (oligoclase–anorthite) have refractive indices greater than Canada balsam (equal to or greater than quartz). Thus for plagioclase feldspars showing extinction angles of less than 20°, a confirmatory Becke test must be done.

Another means of identifying plagioclase feldspars is Schusters method. This is suited to the identification of cleavage flakes which may be seen in smear slides. Cleavage flakes showing slightly better cleavage will show composition planes parallel to cleavage edges, indicating polysynthetic albite law twins with composition planes normal to (001). Schuster's curves (Fig. 36) show the extinction angles corresponding to plagioclase composition. Reference to the curves may enable identification of the plagioclase fragment observed.

Twinning

Feldspars can exhibit at least seven types of twinning and certain types are diagnostic of varieties (Fig. 37). Simple twinning on the Baveno,

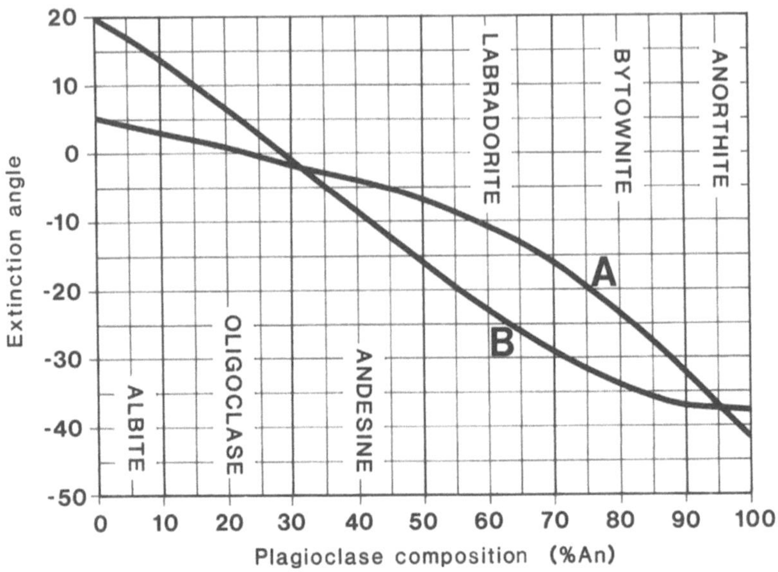

Fig. 36. Schuster's curves showing extinction on {001} (curve A) and {010} cleavages (curve B) (after Kerr, 1977).

Manebach or Carlsbad laws is characteristic of the alkali feldspars. Distinctive polysynthetic twinning in two directions (very close 'grid' twinning) is almost universal in microcline and is diagnostic. Multiple or polysynthetic twinning on the albite law is characteristic of plagioclase. The multiple twin lamellae are straight and parallel. There is a

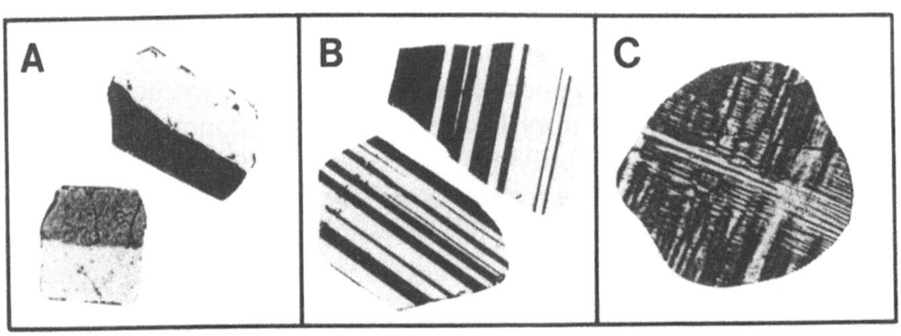

Fig. 37. Diagram showing the different types of twinning commonly seen in different feldspars when viewed in cross-polarised light. (A) Simple Carlsbad twinning which may be seen in all feldspar varieties but is especially common in orthoclase. (B) Multiple or lamellar twinning on the albite law characteristic of the plagioclases. (C) Cross-hatched polysynthetic twinning which is almost universal in microcline and rare in other varieties.

relationship between plagioclase composition and lamellae thickness. They are narrowest in oligoclase and widest in albite and anorthite. Crystal twinning will be best seen in cross-polarised light.

Optical Character

All the feldspars are biaxial and all have large optic angles except sanidine. The alkali feldspars are optically negative while the plagioclase series varies from positive to negative.

Occurrence

Feldspars are ubiquitous in the sediments of the world ocean with plagioclase of intermediate to calcic composition tending to occur in higher concentrations than alkali feldspars (Windom, 1976).

Oceanic Distribution

Feldspars are widely distributed in ocean sediments. In the equatorial Atlantic much of the feldspar may be of aeolian deposition derived from the Sahara (Windom, 1976). In the North and Central Pacific Ocean, Heath (1969) distinguished three distinct mineral associations containing feldspars characterising different geotectonic settings:

1. An acidic or continental association composed of sodic to intermediate plagioclase and alkali feldspars.
2. An island arc association containing pyroxene and plagioclase of intermediate composition.
3. An oceanic association characterised by intermediate to calcic plagioclase and pyroxenes.

Similar associations can probably be recognised in other oceans. Peterson & Goldberg (1962) in an extensive study of feldspar distribution in Pacific sediments concluded that most were derived from within the ocean basin. As yet there are few data on feldspar distribution in the Indian Ocean. Windom (1976) quotes unpublished data indicating a correlation between relatively high amounts of plagioclase and volcanically active areas, e.g. off the coast of Indonesia, Madagascar Rise, etc.

Source

The feldspars in marine sediments originate from both marine volcanism and continental weathering. Those derived from marine volcanism are primarily intermediate to calcic in composition (oligoclase–labradorite) while those of continental provenance are characteristically sodic plagioclases and alkali feldspar (mainly orthoclase and microcline). In general, the feldspar of acid volcanics tends to be sanidine while microcline and orthoclase are characteristic of acid plutonic rocks. Sanidine is particularly characteristic of some sodic volcanic provinces such as the Tyrrhenian Sea. Perthitic feldspars are characteristic of plutonic sources. Feldspars derived from pyroclastics may be broken and may be enclosed in a thin envelope of glass (Pettijohn, 1975). Plagioclase derived from volcanic and hypabyssal rocks may show oscillatory zoning and such zoning is rare in feldspars derived from metamorphic or plutonic sources (Pittman, 1963). Absence of feldspar from a particular sediment may also be of some importance. Fine-grained terrigenous sediments showing high maturation (i.e. lacking feldspar) may suggest that the continental sources were distant and had little relief.

Transport Mechanisms

Continental derived feldspars will pass to the deep sea through fluvial transport, coastal erosion, ice-rafting, mass gravity flows and aeolian deposition. Feldspar derived from within the ocean basin may be redistributed through bottom currents.

Indicator

Although the distribution of feldspars in marine sediments reflects both continental weathering and marine volcanic activity, the similarity of the species supplied by both sources means the distribution of this mineral may yield little information concerning its source or mode of transport. However, as a result of its low stability, sediments containing abundant feldspar are generally uncommon in the deep ocean. Some tuff layers may contain abundant feldspar and deep-sea arkosic sands (generally defined as sands containing over 25% feldspar) of continental derivation may have some palaeoclimatic usefulness. Such sands indicate that either (i) the climatic conditions in the source area were too dry or too cold to allow decomposition of the feldspar or (ii) blockfaulting and rapid tectonic uplift in the source area resulted in rapid erosion and burial of the feldspar before weathering of the grains could be completed (Folk, 1980). Thus deep-sea arkosic sands deposited off cratonic areas can have important palaeoclimatic implications.

For example, Damuth & Fairbridge (1970) inferred ice age aridity in the Amazon River drainage basin from the high percentage of feldspar in redeposited sands of the western equatorial Atlantic.

Main Distinguishing Features

Subhedral prismatic to tabular form, turbidity due to incipient alteration, distinctive twinning (seen in cross-polars): simple in the alkali feldspars, 'grid' twinning in microcline, polysynthetic lamellar twinning in the plagioclases. Low birefringence.

Appearance as Loose Grains

Subhedral prismatic to tabular or rounded grains, white by reflected light. Chipped grains may show planar cleavage planes. May be subtranslucent. Grains have a vitreous lustre.

Appearance in Thin Section

See under 'main distinguishing features'.

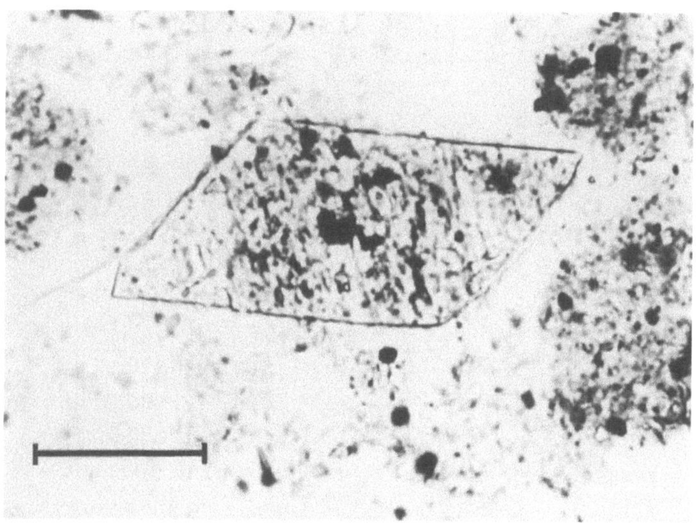

Fig. 38. Euhedral authigenic K-feldspar crystal. Note the trains of inclusions forming ragged lines parallel to the *c* axis which represent previous sections of {110} faces which were subsequently overgrown leaving attached sediment entrapped. Miocene, Southern Columbia Basin, Caribbean Sea (11°05.11′N, 80°22.75′W). DSDP Leg 15, 154-13-1, 80 cm. Plane-polarised light, scale bar equals 0·05 mm. (Courtesy of Dr T. W. Donnelly and the Deep Sea Drilling Project.)

Remarks/Notes

Authigenic feldspars, particularly alkali feldspar, have been reported from several deep-sea sediments, either replacing zeolites (Hay, 1966; and others) plagioclase (Kelts & Mackenzie, 1976; and others) or volcanic glass (Lancelot *et al.*, 1972; Kastner & Siever, 1979). Discrete authigenic alkali feldspar crystals have also been reported from volcaniclastic sediments (Donnelly & Nalli, 1973; Kastner & Gieskes, 1976). Such authigenic feldspar occurs as euhedral and anhedral grains and crystals which usually contain abundant clay inclusions. These often occur as trains of inclusions forming ragged lines parallel to the *c* axis which represent previous sections of {110} faces which were subsequently overgrown leaving attached sediment entrapped (Fig. 38). Authigenic K-feldspar may also occur as formless, multicrystalline aggregates recognisable by their birefringence, relief and texture. In contrast, volcanic alkali feldspar is generally clear.

Feldspars may also be distinguished by staining. Feldspar staining methods are described by Hayes & Klugman (1959) and Bailey & Stevens (1960).

FERROMAGNESIAN MINERALS

For the purpose of smear slide description, the following minerals are considered to fall within this general group:

The amphiboles
The olivines
The pyroxenes

Ferromagnesian minerals can be difficult to identify in smear slides, especially if they occur as anhedral grains. Presence or absence of pleochroism, extinction angle, if cleavage traces or crystal form can be observed, and birefringence are probably the best guides.

Amphiboles

Detrital/volcanic

The amphiboles are a group of inosilicates characterised by a structure of double chains of linked SiO_4 tetrahedra.

The basic general formula is $A_{2-3}B_5Si_8O_{22}(OH)_2$ where A may be Ca, Na or K and B may be Mg, divalent or trivalent Fe, Al or Ti.

Note. The amphiboles comprise a group of minerals with a wide compositional range. Consequently, there may be marked variation in their optical properties, but the following will serve as a general guide.

Form/Shape

The amphiboles are a varied group but they are most commonly seen as subhedral prismatic crystals (Figs 39–41), crystal fragments (Fig. 42), irregular ragged grains, fibrous aggregates (for actinolite–tremolite) and elongate, sometimes platy, cleavage fragments (Fig. 43).

Transparency/Colour in PPL

Usually transparent. Colourless to yellowish-brown or green. Colour and observation of pleochroism (see below) is an important property for distinguishing between different amphiboles in smear slides and thin

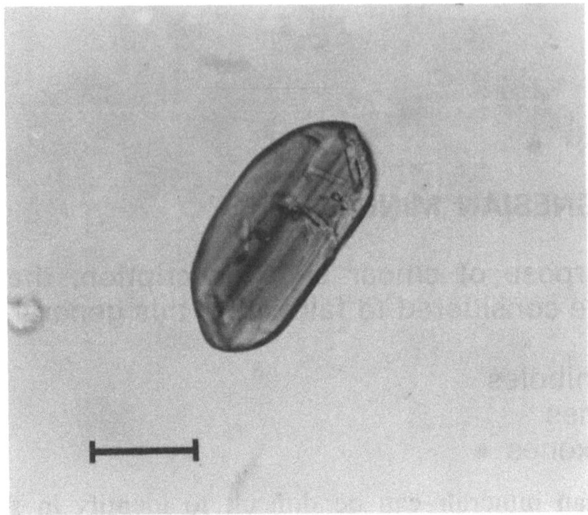

Fig. 39. Rounded hornblende prism, pleochroic in greens. Maastrichtian, southeast flank of Nicaragua Rise, Caribbean Sea (15°52.72′N, 74°36.47′W). DSDP Leg 15, 152-10-1, 147 cm. Plane-polarised light, scale bar equals 0·05 mm.

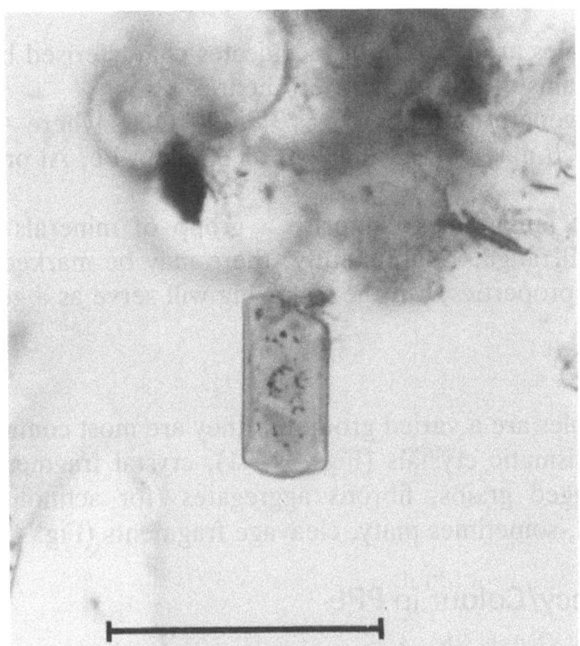

Fig. 40. Euhedral green prismatic hornblende. Pleistocene, Tyrrhenian Sea, Mediterranean. Plane-polarised light, scale bar equals 0·05 mm.

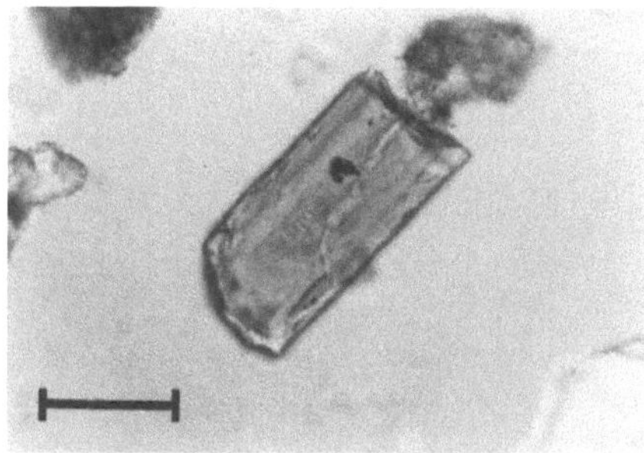

Fig. 41. Subhedral prismatic hornblende, pleochroic in greens. Pleistocene, Tyrrhenian Sea, Mediterranean. Plane-polarised light, scale bar equals 0·05 mm.

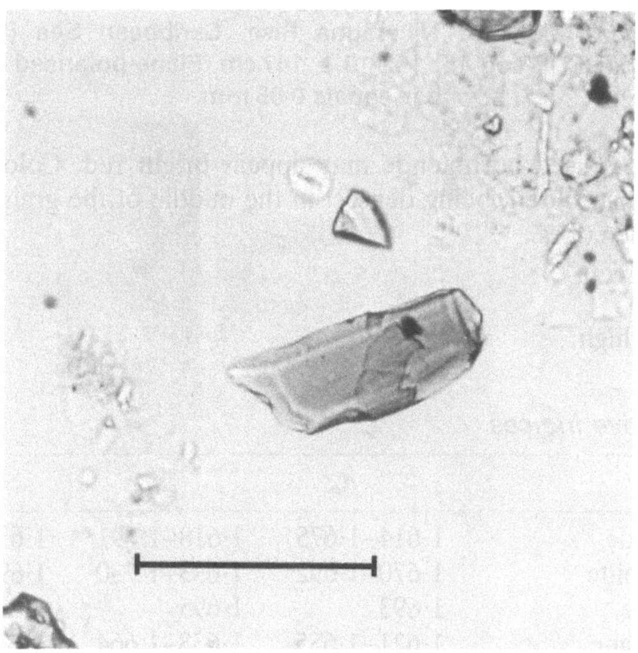

Fig. 42. Broken amphibole crystal, probably hornblende. Green and somewhat pleochroic. Pleistocene, Labrador Sea, Northwest Atlantic (56°47.40′N, 48°19.91′W). DSDP Leg 12, 113-2-3, 7 cm. Plane-polarised light, scale bar 0·05 mm.

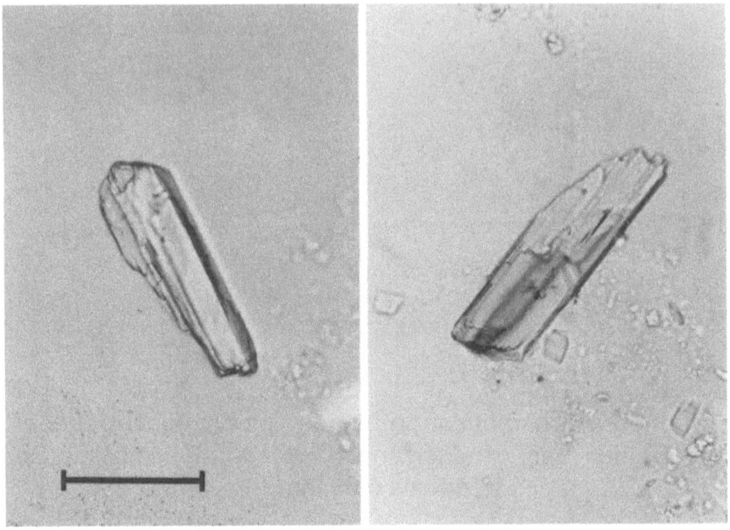

Fig. 43. Prismatic cleavage fragments of green hornblende. Maastrichtian, southeast flank of Nicaragua Rise, Caribbean Sea (15°52.72′N, 74°36.47′W). DSDP Leg 15, 152-10-1, 147 cm. Plane-polarised light, scale bar equals 0·05 mm.

sections. Oxidised hornblende may appear bright red. Colour may be unevenly distributed, being densest in the middle of the grain and paler towards the edges.

Relief

Generally high.

Refractive indices

	n_α	n_β	n_γ
Hornblende	1·614–1·675	1·618–1·691	1·633–1·701
Lamprobolite	1·670–1·692	1·683–1·730	1·693–1·760
Riebeckite	1·693	1·695	1·697
Glaucophane	1·621–1·655	1·638–1·664	1·639–1·668
Actinolite–tremolite	1·600–1·628	1·613–1·644	1·625–1·655

Birefringence

Moderate (0·01–0·03). In cross-polars, interference colours will be bright, but these may be masked or modified by body colour.

Alteration

Minerals of the amphibole group are apparently relatively stable in the marine environment, although some amphiboles (e.g. hornblende and actinolite–tremolite) may show alteration to chloritic matter. Downhole amphibole grains may show etching through intrastratal solution, which may result in the grains having 'hacksaw terminations' (Fig. 44). Intrastratal solution of ferromagnesian minerals will result in corroded and pitted grain surfaces and skeletal grains (Edelman & Doeglas, 1932). However, not all ragged 'hacksaw' grains are necessarily produced through intrastratal solution. Breakage during transport can produce irregular shapes and some grains may owe their ragged appearance to their original shapes in the source rocks (Hubert, 1971).

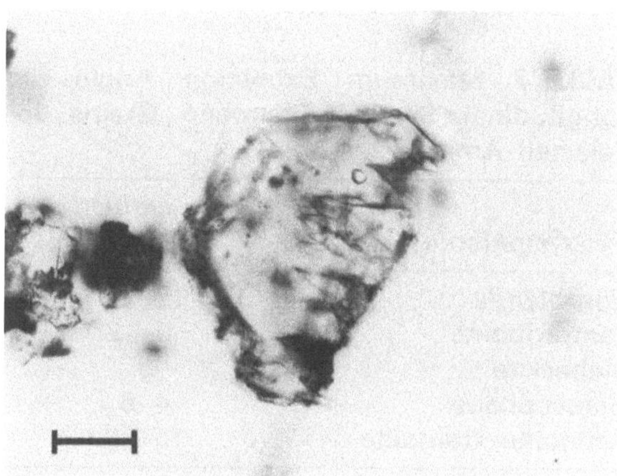

Fig. 44. Ferromagnesian grain showing 'hacksaw' terminations due to intrastratal solution. Upper Eocene, Kings Trough, Northeast Atlantic (44°02.12′N, 21°48.79′W). Plane-polarised light, scale bar equals 0·05 mm.

Crystal System

Nearly all amphiboles are monoclinic.

Cleavage

All amphiboles show perfect {110} cleavage in two directions intersecting at angles of about 55° and 125°.

Pleochroism

Common hornblende is strongly pleochroic in shades of brown, green, yellow and bluish-green; basaltic hornblende (lamprobolite) is pleochroic from light yellow to dark red-brown. Riebeckite, a sodic amphibole, is pleochroic from almost opaque to dark inky blue and pale yellow. Glaucophane, another sodic amphibole, is pleochroic in shades of blue and violet.

Extinction Angle

The maximum extinction angle on {010} varies from 0° to 30° depending on the amphibole. However, individual minerals within the amphibole group may show a range of extinction angle values (Table 7). Cross-sections will show symmetrical extinction but such sections are unlikely to be seen in smear slides.

TABLE 7 Maximum Extinction Angles in Longitudinal Sections/Prismatic Grains for Selected Amphiboles

Amphibole	*Max. extinction angle (°)*
Hornblende	12–30
Lamprobolite	0–12
Riebeckite	5
Glaucophane	4–6
Actinolite–tremolite	10–20

Data from Kerr (1977).

Twinning

Twins with {100} as the twin plane are rather common in hornblende.

Optical Character

Most common amphiboles give biaxial negative interference figures.

Occurrence

Amphiboles appear relatively common in most marine sediments and can originate from both terrestrial and oceanic sources. Green hornblende is probably the most common amphibole occurring as a detrital mineral. Basaltic hornblende (lamprobolite), which is usually brown in colour, may occur in pyroclastics but is generally rare. Riebeckite and actinolite–tremolite are also rare in detrital sediments. Glaucophane may be locally common in marine sediments adjacent to high pressure regional metamorphic belts.

Oceanic Distribution

Biscaye (1965) in a detailed study of amphibole distribution in Atlantic Ocean sediments found its distribution pattern correlated with that of chlorite. From this he surmised that it had a predominantly continental origin. Windom (1969) found amphiboles to be common constituents of North Pacific sediments and also of dusts in adjacent continental snowfields. From this he deduced that much of the amphibole was of continental derivation and that aeolian transport was particularly important in its distribution. There is, as yet, little data on amphibole distribution in the Indian Ocean.

Transport Mechanism

Both Biscaye's (1965) and Windom's (1969) studies suggest that most of the amphiboles in marine sediments are derived from continental sources, and that aeolian deposition of fine-grained amphibole may be an important transport mechanism, especially in equatorial regions. Other transport mechanisms from the continents to deep ocean will include fluvial transport, coastal erosion, ice-rafting and mass gravity flows.

Source

Common hornblende is commonly derived from intermediate felsic igneous rocks while blue-green hornblende is considered to indicate a metamorphic source (Pettijohn, 1975). Basaltic hornblende (lamprobolite) indicates a volcanic source. Riebeckite is derived from sodic granites, syenites and trachytes. It is often associated with aegirine. Actinolite–tremolite usually has an origin in metamorphic rocks or as an alteration product of other ferromagnesian minerals in igneous rocks. Glaucophane is usually derived from schists and gneisses occurring in high pressure regional metamorphic belts.

Main Distinguishing Features

Prismatic habit, {110} cleavage in two directions at angles of about 55° and 125°, colour, pleochroism in some forms, high relief, low extinction angle.

Appearance as Loose Grains

Rounded or long or short prismatic grains, varying from light green to black in colour. Broken grains may show intersecting cleavage planes, the intersections of which are marked by an obtuse angle. Grains will have a vitreous lustre.

Appearance in Thin Section

Euhedra will show characteristic rhombic to pseudohexagonal cross-sections. Crystals will be of long or short prismatic habit or occur as asbestiform aggregates. The amphiboles usually show bright interference colours up to the second order but these are often masked or modified by the body colour.

Olivine (Syn. Chrysolite)

Detrital/volcanic

The olivines form a solid solution series with the general formula A''_2SiO_4 (where A'' may be Mg, divalent Fe, Mn, Ca (part)). The end members of the solid solution series are forsterite (Mg_2SiO_4) and fayalite (Fe_2SiO_4).

Form/Shape

When olivine is present in marine sediments, it usually occurs as subangular, often fractured anhedra. Well-formed crystals are rare. Grains may show alteration to fine-grained decomposition products, especially along crystal fractures.

Fragmentation Characteristics

Olivine shows conchoidal fracture, thus crystal fragments are likely to be irregular.

Grain Size Range

Silt–sand size range.

Transparency/Colour in PPL

Usually transparent to colourless. Serpentinised grains may be greenish and slightly turbid.

Relief

Fairly high, $n_\alpha = 1{\cdot}651–1{\cdot}681$, $n_\beta = 1{\cdot}670–1{\cdot}706$, $n_\gamma = 1{\cdot}689–1{\cdot}718$.

Birefringence

Strong (0·037–0·041), thus grains will show vivid interference colours.

Alteration

Olivine is unstable in the marine environment and rapidly alters to serpentine minerals.

Crystal System

Orthorhombic.

Cleavage

Imperfect parallel to {010} but cleavage will be obscure in smear slides.

Pleochroism

Absent.

Extinction Angle

In euhedral crystals, extinction is parallel to crystal outlines and cleavage traces. This will rarely be a measurable property in smear slides.

Twinning

Twinning in olivine crystals is occasionally seen but the twin lamellae are broad and not well-defined.

Optical Character

Olivine usually shows a biaxial positive interference figure with a large axial angle, although olivine high in iron is optically negative. The size of the axial angle may vary within single crystals reflecting zoning.

Associated Minerals

Olivine is frequently associated with calcic plagioclase and pyroxenes.

Occurrence

Olivine is unstable in sedimentary environments and is rare in marine sediments. However, it may be locally common in sediments close to active volcanic centres. Olivine and pyroxenes may occur as discrete spherules up to 200 μm across and these are probably of extraterrestrial origin (Arrhenius, 1963). These are only of minor importance in marine sediments.

Oceanic Distribution

Olivine will be largely restricted to actively volcanic regions.

Indicator

Proximity of active volcanism.

Main Distinguishing Features

High relief, strong birefringence, mineral association. However, olivine is very difficult to identify positively in smear slides. Its low stability means it will be rarely seen except in basic volcaniclastic sediments. In these, the type of assemblage will alert the geologist to its likely presence.

Appearance in Thin Section

Olivine is an important rock-forming mineral and will be seen as polygonal anhedra and euhedral phenocrysts of distinctive outline. Extinction will be parallel to crystal outlines. Maximum interference colours at 30 μm will be upper second order.

Pyroxenes

Detrital/volcanic

The pyroxenes comprise a group of rock-forming silicates having a chain structure of linked SiO_4 tetrahedra with a unit form of (SiO_3): Si_2O_6 and Si_4O_{12} being the most commonly quoted.

The general formula may be written $\underset{\sim}{A}_2Si_2O_6$ (or $\underset{\sim}{A}SiO_3$ in some cases) where $\underset{\sim}{A}$ is Mg, divalent Fe or Ca.

Like the amphiboles, to which they form a more or less parallel group, the pyroxenes comprise a group of minerals with a wide compositional range. Consequently, there may be a marked variation in their optical properties, but the following will serve as a general guide.

Form/Shape

A varied group but most commonly seen as corroded stout prismatic crystals, irregular anhedra and cleavage fragments (Figs 45 and 46). Grains may be pitted and etched and contain small inclusions.

Grain Size Range

In the silt–sand size range.

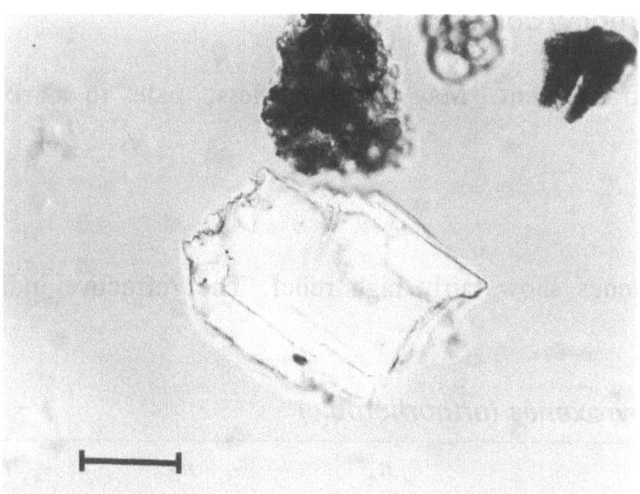

Fig. 45. Prismatic fragment of orthopyroxene. Pleistocene, Tyrrhenian Sea, Mediterranean. Plane-polarised light, scale bar equals 0·05 mm.

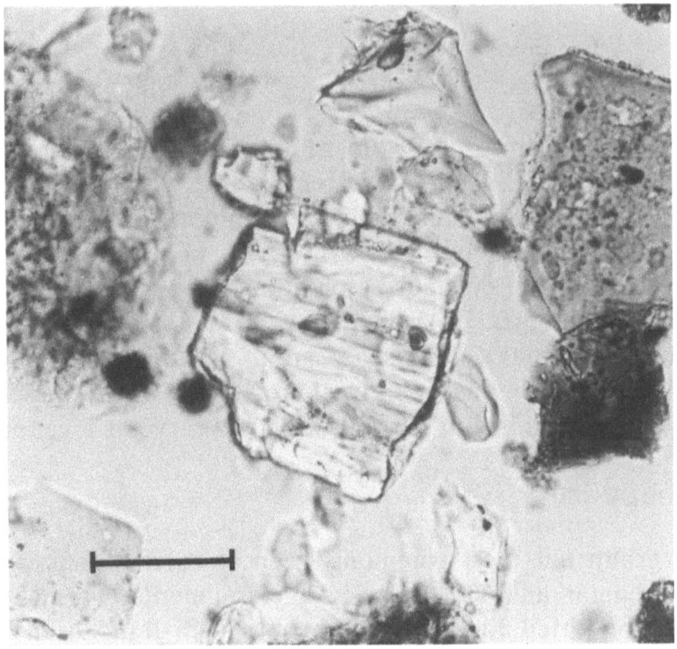

Fig. 46. Subhedral clinopyroxene grain. Cleavage traces are just visible. Pleistocene, Tyrrhenian Sea, Mediterranean. Plane-polarised light, scale bar equals 0·05 mm.

Transparency/Colour in PPL

Usually transparent. May be colourless, pale to dark green or brownish.

Relief

All pyroxenes show fairly high relief. The refractive indices are as follows:

Orthopyroxenes (orthorhombic)

	n_α	n_β	n_γ
Enstatite	1·650–1·665	1·653–1·670	1·658–1·674
Hypersthene	1·673–1·715	1·678–1·728	1·683–1·731

Clinopyroxenes (monoclinic)

	n_α	n_β	n_γ
Diopside	1·650–1·698	1·657–1·706	1·681–1·727
Augite	1·688–1·712	1·701–1·717	1·713–1·737
Pigeonite	1·680–1·718	1·698–1·725	1·719–1·744
Hedenbergite	1·732–1·739	1·737–1·745	1·751–1·757
Aegirine–Augite	1·680–1·745	1·687–1·770	1·709–1·782
Aegirine	1·745–1·777	1·770–1·823	1·782–1·836

Birefringence

The monoclinic pyroxenes are distinguished by their moderate birefringence (about 0·025). The orthorhombic pyroxenes have low birefringence (about 0·010), thus giving lower order interference colours.

Alteration

Most pyroxenes are relatively stable in the marine environment and will appear unaltered. However, downhole pyroxenes, like amphiboles, may show crystallographically-controlled etch patterns, sometimes giving characteristic saw-tooth terminations to the crystals.

Crystal System

Most pyroxenes are monoclinic (termed clinopyroxenes), except the enstatite–hypersthene–ferrosilite series which is orthorhombic (orthopyroxenes).

Cleavage

All pyroxenes have a perfect cleavage parallel to {110} which gives two cleavage directions meeting at almost 90°.

Pleochroism

Hypersthene may be pleochroic from greenish to pale red, aegirine–augite is pleochroic from yellow-green to greenish and aegirine is strongly pleochroic from yellow to dark green.

Extinction Angle

Orthopyroxenes show straight extinction while clinopyroxenes have oblique extinction (measured from the *c* axis in the plane {010}). The extinction angles for clinopyroxenes vary from 6 to 50° depending on the pyroxene (Fig. 47).

Twinning

Twins with {100} as twin planes are common in augite, diopside, and aegirine–augite. Polysynthetic twinning wtih {100} as the twin plane is characteristic of pigeonite.

Optical Character

Most pyroxenes have biaxial positive interference figures with large axial angles, exceptions are pigeonite, which is biaxial positive with a

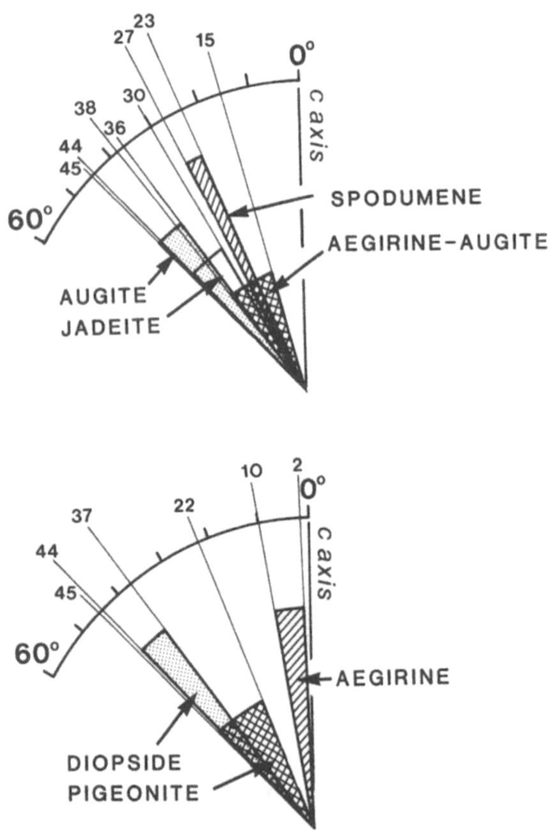

Fig. 47. Clinopyroxene extinction angles on {010} (data from Kerr, 1977).

small axial angle, and aegirine and hypersthene, which are both biaxial negative with large axial angles.

Occurrence

Pyroxenes appear to be relatively common in marine sediments.

Oceanic Distribution

Pyroxenes, notably pigeonite, are an important constituent of oceanic tholeiites. Consequently, pyroxenes will be relatively common sedimentary components in oceanic volcanic provinces. Aegirine-augite is particularly common and easily identified. Few studies have been made of pyroxene distribution in the world ocean but Goldberg & Griffin (1964) found a latitudinal trend in the abundance of pyroxene in Atlantic Ocean sediments. They found an increase in pyroxene concentration at higher latitudes, perhaps reflecting larger amounts of volcanic material in these regions.

Source/Transport Mechanisms

Pyroxenes in marine sediments will be derived from both oceanic and continental sources. Transport from continental sources may be by fluvial transport, coastal erosion, ice-rafting, and mass gravity flows. Pyroxenes derived from within the ocean basins may be concentrated with other heavy minerals by bottom currents. Most pyroxenes will be derived from mafic igneous rocks. Aegirine–augite and aegirine are characteristic of sodic igneous rocks such as trachytes, syenites and phonolites.

Indicator

Although pyroxenes tend to be ubiquitous in deep-sea sediments, different types may be useful indicators of the origin of pyroclastics.

Main Distinguishing Features

Pyroxenes are most likely to be confused with amphiboles in smear slides. In thin section, the two mineral groups may be distinguished by their cleavage but this requires examination of basal sections which are unlikely to be well-presented in smear slides. However, amphiboles are most commonly coloured and pleochroic and thus are only likely to be confused with sodic pyroxenes such as aegirine. The extinction angle on the trace of the cleavage in grains elongated parallel to the *c* axis is also a useful guide, being up to 50° for clinopyroxenes and 20° or less for amphiboles. Orthopyroxenes are distinguished by their straight extinction and silt-sized grains will show low interference colours.

Appearance as Loose Grains

Rounded anhedra or stout prismatic eu/subhedra, usually dark greenish-black in colour. Broken grains may show intersecting planar cleavage planes, intersecting at 90°. Grains will be transparent to opaque and have a vitreous lustre.

Appearance in Thin Section

Plane-polarised light: high relief, stout prismatic form, two cleavage directions meeting at almost 90°. Most pyroxenes are almost colourless to neutral in colour; aegirine is dark green and pleochroic and hypersthene may have pink-green pleochroism. Cross-polarised light: *clinopyroxenes,* moderate birefringence giving second order colours. Oblique extinction in sections cut approximately parallel to the *c* axis. *Orthopyroxenes,* low birefringence giving first order colours. Straight extinction.

GLAUCONITE

Authigenic/detrital

Note. Greenish sand-sized grains described as 'glauconite' are usually mixtures of minerals rather than single mineral species (Burst, 1958*a*; 1958*b*), although a constituent common in such grains is the mineral glauconite, a hydrated iron-rich micaeous clay mineral of the illite group (K Mg (Fe, Al)$(SiO_3)_6.3H_2O$).

Form/Shape

Generally smooth grains of ovoid and/or lobate form (Figs 48–50) but a wide range of grain shapes may occur including tabular, discoidal, mammillary, capsule-shaped, composite pellets, vermicular grain clasts

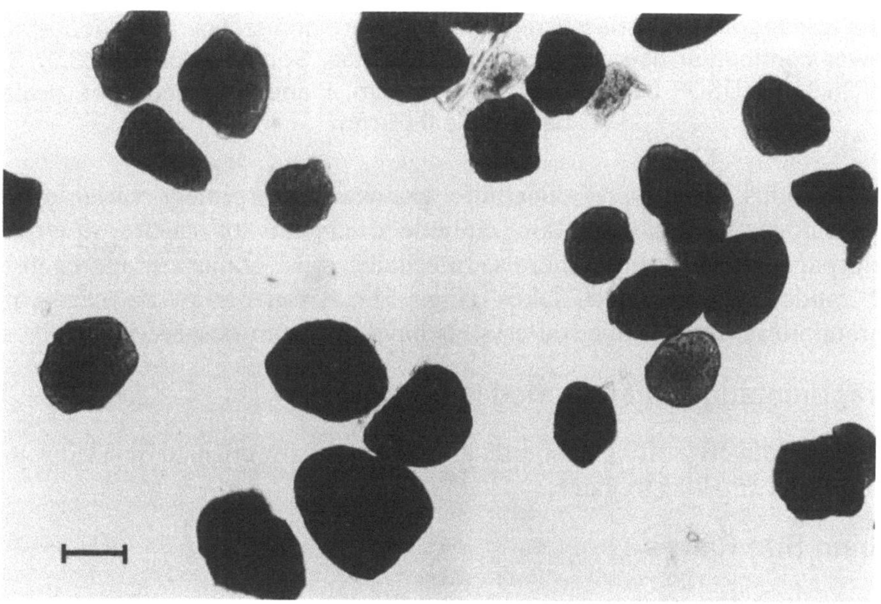

Fig. 48. Generally smooth rounded grains of glauconite, dark green in colour. Quaternary, Feni Drift, Rockall Trough, Northeast Atlantic (53°13.46′N, 18°53.69′W). DSDP Leg 94, 610-4-5, 139–140 cm (picked fraction). Plane-polarised light, scale bar equals 0·05 mm.

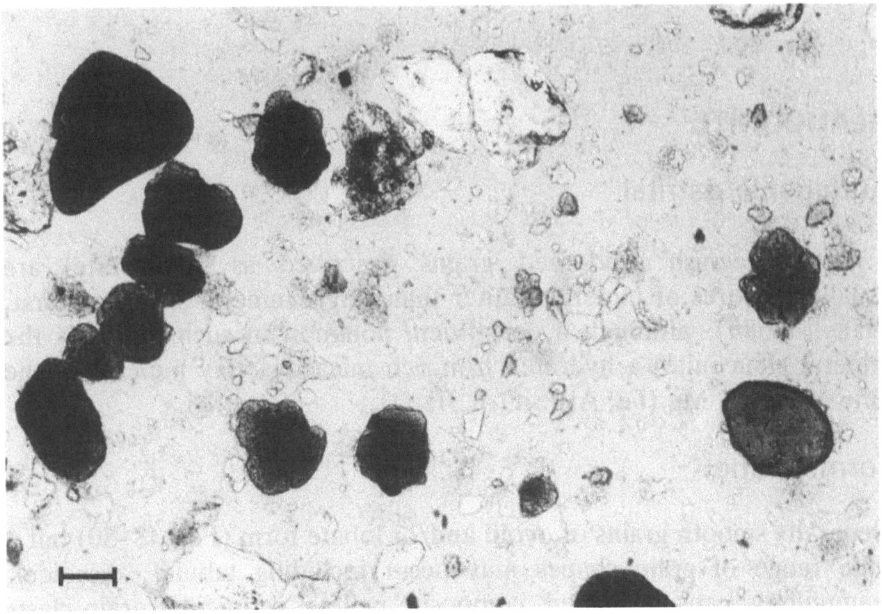

Fig. 49. Irregular and smooth glauconite grains, bright green in colour. The transparent colourless irregular grains are quartz. Lower Palaeocene, lower continental rise, west of Cape Agulhas, South Africa (35°03.97′S, 15°26.91′E). DSDP Leg 40, 361-10-2, 129 cm. Plane-polarised light, scale bar equals 0·05 mm.

and moulds. However, glauconite grains are extremely variable in mineralogy and may contain embedded crystals of calcite, quartz, feldspar and other clay minerals. Internally, most grains are aggregates of randomly orientated flakes (Fig. 51). Grains may be casts of foraminiferal tests. Euhedral crystals have not been observed.

Fragmentation Characteristics

Being relatively soft, glauconite grains are easily crushed resulting in clusters of small aggregates.

Grain Size Range

Typically in the sand-size range, generally 100–500 μm in size.

Transparency/Colour in PPL

Transparency and colour will depend on the size and thickness of the grain. Relatively thin or small grains may be semi-translucent and vivid

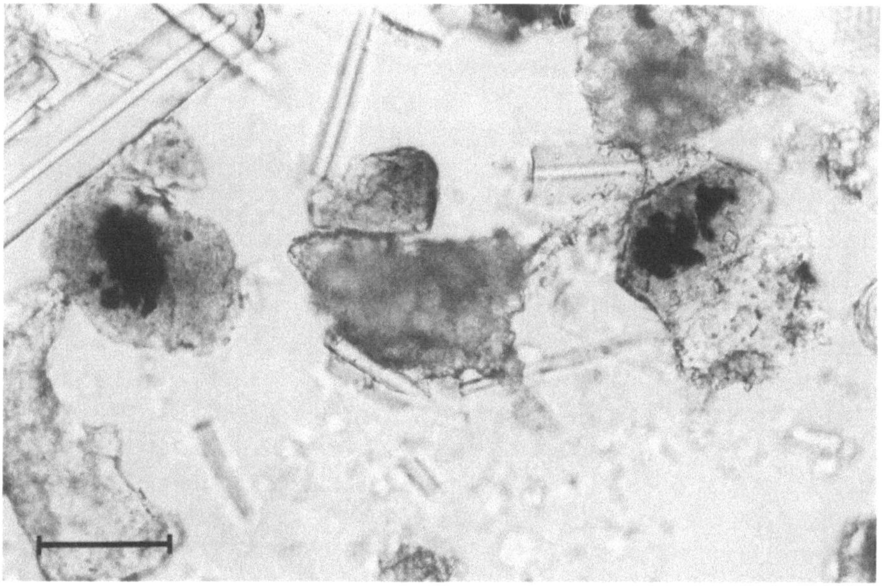

Fig. 50. Green glauconite with opaque cores (probably pyrite). The transparent elongate fragments, sometimes with a central hollow tube, are broken sponge spicules (biogenic silica). Middle Eocene, western flank, Bermuda Rise, Northwest Atlantic (30°50.39′N, 67°38.86′W). DSDP Leg 1, 6-4-3, 57 cm. Plane-polarised light, scale bar equals 0·05 mm.

green, however, colour may vary from dark green, almost black to a pale greenish-yellow. The shade of green is, in part, related to the relative amounts of iron and aluminium present (Burt, 1932; Borchert & Braun, 1963) and particularly to the ratio of ferrous–ferric iron (Weyl, 1951; Keller, 1953; 1958). Large grains may appear black and opaque but such grains frequently have translucent green margins.

Birefringence

Although the birefringence shown by the mineral glauconite is moderate to rather strong (0·020–0·032) most grains show aggregate polarisation with low order interference colours, which may be masked by the colour of the mineral.

Alteration

Oxidation of glauconite will cause the colour of the outer skin to change to a rusty brown. Such grains may be reported as limonite.

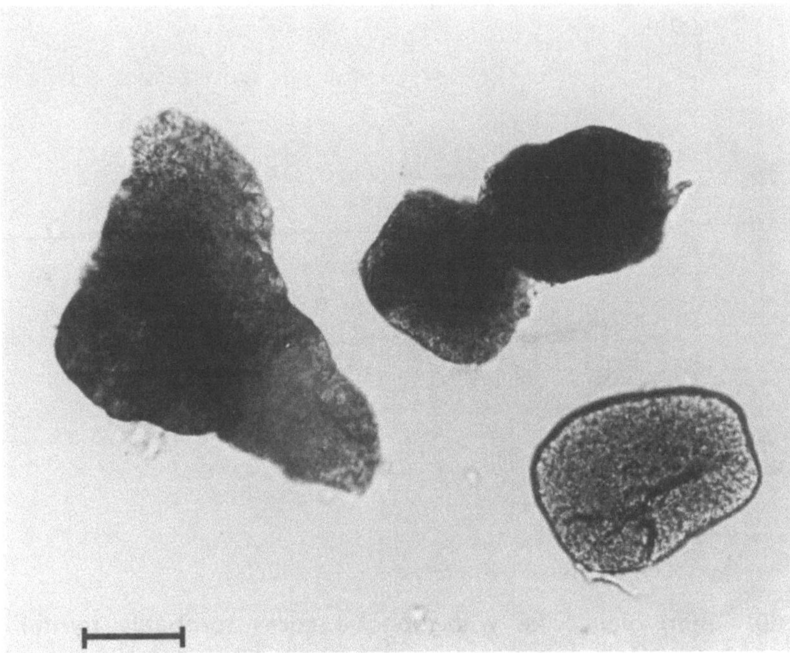

Fig. 51. Glauconite grains. Detail of the internal aggregate structure may be seen in the thinner grain (bottom right). Quaternary, Feni Drift, Rockall Trough, Northeast Atlantic (53°13.46′N, 18°53.69′W). DSDP Leg 94, 610-4-5, 139–140 cm (picked fraction). Plane-polarised light, scale bar equals 0·05 mm.

Note. The following optical data specifically refers to the mineral glauconite. The aggregate polarisation shown by most grains renders detailed optical work impossible and the reported optical data refers only to grains with vermicular or orientated microcystalline structure.

Crystal System

Monoclinic (?).

Cleavage

The mineral glauconite has perfect cleavage in one direction {001}.

Pleochroism

The mineral glauconite may be pleochroic from yellow to green, although this will rarely be apparent in smear slides.

Extinction

The mineral glauconite extinguishes practically parallel to the cleavage.

Twinning

None recorded.

Optical Character

Cleavage flakes give a biaxial negative figure with a small axial angle.

Group Variability

The optical and other properties of glauconite grains will vary according to mineralogical composition, degree of drying, presence of impurities and degree of surface oxidation.

Associated Minerals

Glauconite has been reported as being associated with a wide range of minerals and deposit types including phosphatic deposits and sedimentary iron ore.

Occurrence/Sedimentary Environments

Although glauconite is most commonly found in marine sediments it has also been reported from lacustrine and other alluvial occurrences. Glauconite found in sediments may be authigenic or detrital. The distinction between these usually involves consideration of (1) the petrology of the rock or sediment, (2) morphology of the grains and (3) mineralogy of the grains and associated sediment. Glauconite forms mainly on the continental shelves away from large estuaries. Its formation is governed largely by depth, temperature and oxygenation (McRae, 1972). Glauconite is considered unlikely to form in water shallower than about 15 m due to wave-induced turbulence (Cloud, 1955) and it is commonly held that glauconite will be rare in water deeper than the shelf margin (500 m) (Fairbridge, 1967). Therefore it would appear that most occurrences in the deep sea will be of detrital glauconite and not authigenic glauconite formed at these depths, although there is still some dispute on this (see for example, Odin & Stephan, 1981). Glauconite is most widespread today in warm shelf seas

(15°C min., 20°C max.) (Takahashi & Yagi, 1929; Porrenga, 1967). From its content of both ferric and ferrous iron, it seems that glauconite forms under moderately reducing conditions (perhaps in microenvironments provided by decaying organisms) and it is thought that the formation of glauconite is facilitated by the presence of decaying organic matter (Cloud, 1955; and others). In the geological record, glauconite occurs widely in sedimentary rocks of virtually all ages.

Oceanic Distribution

Glauconite may be found today in all the oceans of the world, although authigenic glauconite is largely formed on the continental shelves. Logvinenko (1982) notes a correlation between the distribution of detrital glauconite and areas of glauconite-bearing rocks on land and states a negative correlation with climate and bathymetry.

Source/Transport Mechanisms

Glauconite suggests a source characterised by open marine waters, nearby continental shores (with a probable water depth of 15–500 m) and the presence of biological activity. Turbidity currents and other forms of sediment gravity flows originating in shelf/outer slope depths are probably mainly responsible for the transport of glauconite into the deep ocean.

Origin (Formation of the mineral glauconite)

The origin of the mineral glauconite is still a matter of debate. The 'layer lattice theory' (Burst, 1958*a*; 1958*b*) holds that the formation of glauconite requires (1) a degraded layer silicate lattice, (2) a plentiful supply of Fe and K, and (3) a suitable redox potential (e.g. produced by decaying organic matter, probably in microenvironments, even in generally oxidising macroenvironments). The process of glauconitisation consists of absorption of Fe and K by the degraded lattice leading to a reduction in the amount of expandable layers so that, given time and suitable chemical conditions, the material approaches the composition of glauconite *senso stricto*.

Formation of morphological glauconite: (i.e. aggregates of fine-grained glauconite mixed with other minerals) may result from:

(1) the alteration of organic matter in macro- and microfossils;
(2) alteration of faecal pellets;

(3) transformation of biotite;
(4) agglomeration of clay-sized particles;
(5) replacement processes; or
(6) precipitation on or alteration of mineral surfaces to give pigmentary glauconite.

Diagenesis

Glauconite survives deep burial and appears to have long term stability. However, Dapples (1967) reports its instability during folding and its reaction to more intense metamorphism is unknown.

Indicator/Uses

Glauconite suggests an origin in open marine waters on the continental shelves. However, the rather wide range of conditions possible for glauconite formation (Cloud, 1955) and its common detrital occurrence makes it less useful as an indicator of particular environments than has frequently been proposed. However, authigenic glauconite indicates low sedimentation rates as the mineral forms most readily at the water/sediment interface and rapid sedimentation would arrest the glauconitisation process by burying the developing grains. Glauconitic horizons may be useful for correlation (Burst, 1958*a*) and even the presence or absence of glauconite may be sufficient for correlation (Triplehorn, 1966). If the grains are syngenetic with the deposit, glauconite may be used as a basis for K–Ar (Dalrymple & Lamphere, 1970) and Rb–Sr dating (Herzog *et al.*, 1958; Hurley *et al.*, 1960).

Main Distinguishing Features

Form, shape, colour, size range, fragmentation characteristics, aggregate polarisation.

Appearance as Loose Grains

Ovoid and/or lobate smooth grains, in the sand-size range (0·06–2·00 mm) of varying shades of green. Low hardness (2·0 on the Moh scale). A wide range of grain shapes may occur.

Appearance in Thin Section

Plane-polarised light: green, yellow-green or olive-green grains or pellets that are in part aggregates of minute crystals and partly single crystals. May be pleochroic from yellow to green. Grains are often casts of foraminiferal tests. In cross-polars: aggregate polarisation common, moderate birefringence (first to second order colours) which may be masked by body colour.

GYPSUM (and anhydrite)

Detrital and authigenic
$CaSO_4.2H_2O$
Hydrous calcium sulphate

Form/Shape

Typically occurs as euhedral to subhedral tabular crystals of simple habit (Figs 52 and 53) or platy cleavage fragments (Fig. 54), although it may occur as irregular anhedra. Large crystals may develop a lenticular habit through rounding of the prism {111} and pinacoid {103} (Siesser & Rogers, 1976). Cruciform interpenetration twins of diametral prisms may occur but are generally rare. Swallowtail contact twins of delicate tabular selenite may occur. Gypsum may show a wide variety of form and may occur as radiating clusters or rosettes of bladed and/or acicular crystals.

Fragmentation Characteristics

Fragments cleaved on {010} will yield thin folia but cleavage along {100} and {011} will give rhomb-shaped fragments.

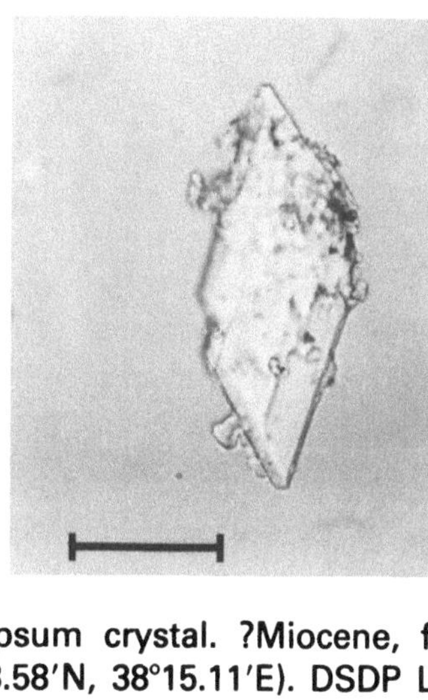

Fig. 52. Euhedral gypsum crystal. ?Miocene, from east of Atlantis II Deep, Red Sea (21°18.58′N, 38°15.11′E). DSDP Leg 23, 225-25-CC (core-catcher). Plane-polarised light, scale bar equals 0·05 mm.

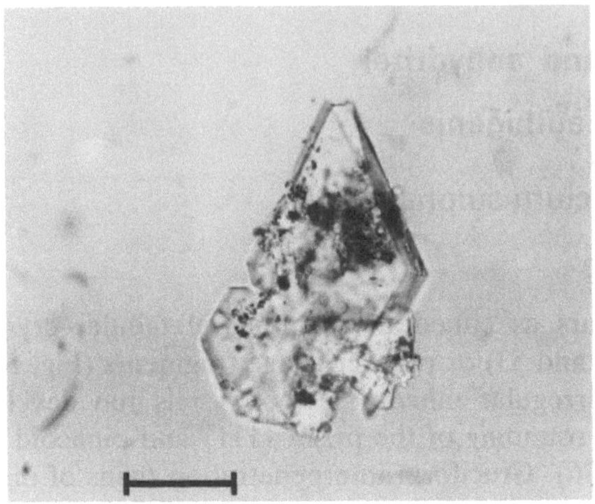

Fig. 53. Gypsum crystal. ?Miocene, from east of Atlantis II Deep, Red Sea (21°18.58'N, 38°15.11'E). DSDP Leg 23, 225-25-CC (core-catcher). Plane-polarised light, scale bar equals 0·05 mm.

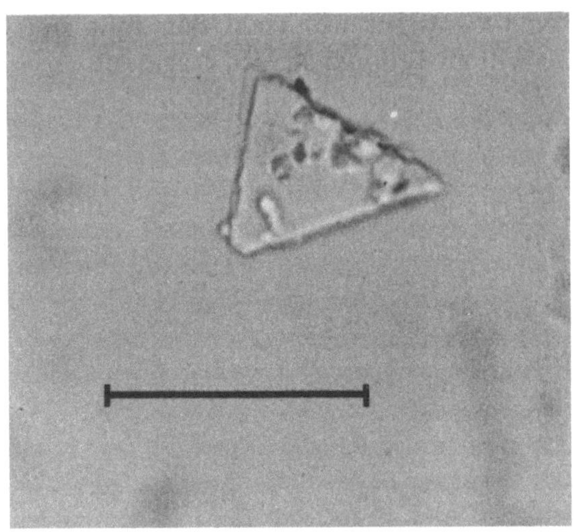

Fig. 54. Gypsum cleavage fragment. ?Miocene, from east of Atlantis II Deep, Red Sea (21°18.58'N, 38°15.11'E). DSDP Leg 23, 225-25-CC (core-catcher). Plane-polarised light, scale bar equals 0·05 mm.

Grain-Size Range

Crystals show a wide range of grain size but generally fall in the range 10 to 300 μm.

Transparency/Colour in PPL

Most of the gypsum seen in smear slides is the colourless transparent crystalline variety, selenite. Authigenic selenite may have variable transparency due to included clay.

Relief

Low. Refractive indices: $n_\alpha = 1{\cdot}520$, $n_\beta = 1{\cdot}522$, $n_\gamma = 1{\cdot}529$.

Birefringence

Rather weak (0·009), about the same as that of quartz. Therefore, even relatively thick grains will show relatively low order interference colours under cross-polars.

Crystal System

Monoclinic.

Cleavage

Gypsum has three unequal cleavages: {010} perfect, {100} distinct, $\{\bar{1}11\}$ distinct.

Pleochroism

Absent.

Extinction Angle

Parallel to {010} cleavage.

Twinning

Swallowtail twins twinned on {100} is the commonest twin type. Heating the slide may result in polysynthetic twinning (Kerr, 1977).

Optical Character

The interference figure is biaxial positive with a moderate axial angle.

Associated Minerals

Authigenic gypsum in deep-sea sediments is frequently associated with pyrite and organic debris (e.g. worm tubes, foraminifera tests). Rounded detrital gypsum may be associated with clastic minerals and turbiditic detritus representing derivation from a nearshore environment by turbidity currents.

Occurrence/Source

In the deep sea, gypsum can occur either as a detrital mineral, ultimately derived from ancient or recent evaporitic deposits and transported to the deep sea by mass gravity flows, or as an authigenic mineral, commonly associated with reducing conditions (Criddle, 1974; Siesser & Rogers, 1976; Briskin & Schreiber, 1978). These writers regard such authigenic gypsum formation to occur via the reaction of calcium derived from the dissolution of biogenic material with sulphate derived by diffusion from overlying waters. However, Schnitker *et al.* (1980) suggest that authigenic gypsum may form as a result of iron sulphide accumulation in carbonate sediments followed by the intrusion of oxygen into the sediment, perhaps through bioturbation or through the replacement of stagnant bottom waters by oxygenated water. The frequent association of authigenic gypsum with pyrite lends support to this model.

Detrital gypsum may be distinguished from authigenic gypsum in that it is usually subhedral or anhedral, cleaved and abraded. Authigenic gypsum occurs as euhedral, unbroken crystals which contain abundant clay inclusions.

Xavier & Klemm (1979) have reported the occurrence of small authigenic (15 μm) twinned and untwinned euhedra and intergrown rosettes of gypsum within manganese nodules from the Central Pacific. These have presumably formed through diffusion processes within the nodules leading to the build-up of Ca and SO_4 ions to the point of precipitation.

Oceanic Distribution

Masssive gypsum beds, some of which appear to have replaced algal mats (Nesteroff, 1973) were cored within the Messinian evaporite sequence in the Tyrrhenian and Ionian basins during DSDP Leg 13 (Ryan *et al.*, 1973). These now lie below deep marine sediments and were deposited during the Messinian desiccation event (Nesteroff, 1973).

Reworked gypsum of continental derivation, presumably transported downslope and quickly buried to prevent dissolution, has been noted at DSDP site 165 in the Pacific (Winterer *et al.*, 1973), at sites 232, 236 and 238 in the Indian Ocean (Cronan *et al.*, 1974) and by Briskin & Schreiber (1978) in piston cores from the South Atlantic. Authigenic gypsum formation may be a fairly common occurrence, albeit ephemeral. Briskin & Schreiber (1978) distinguish two distinct types of authigenic gypsum:

1. Tabular gypsum euhedra, which have been described from the Mediterranean (Robert & Chamley, 1974), off the coast of SW Africa (Siesser & Rogers, 1976) and from the South Atlantic (Briskin & Schreiber, 1978).
2. Prismatic or acicular gypsum. Briskin & Schreiber (1978) describe organically rich clays containing abundant gypsum needles (62–80 μm in length) from the Argentine Basin continental slope. These needles appeared particularly common in burrow infills which sometimes comprised 80% gypsum.

Main Distinguishing Features

Characteristic euhedral to subhedral form, low relief, weak birefringence.

Appearance as Loose Grains

Transparent to translucent euhedral to subhedral colourless grains which have a vitreous lustre. Grains are soft (hardness 2·0 on the Moh scale) and can be crushed and broken under pressure from a needle. Complete euhedral crystals have a distinctive diamond shape.

Appearance in Thin Section

Usually occurs as anhedral to subhedral aggregates which sometimes show a fibrous structure. Colourless in PPL and with low relief. Weak birefringence giving interference colours of the first order and extinction parallel to the {010} cleavage.

Notes/Remarks

Care is needed in the describing of older cores as gypsum may precipitate upon desiccation of interstitial solutions during core storage (Olausson, 1961; Arrhenius, 1963).

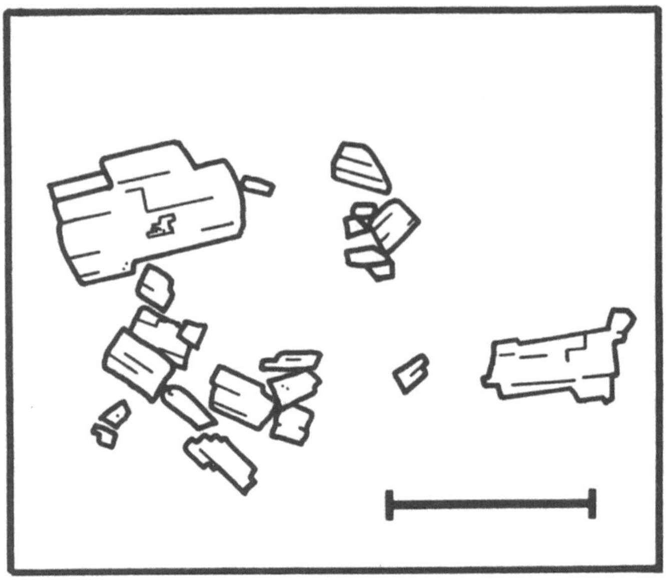

Fig. 55. Subhedral anhydrite crystals. ?Miocene, east of Atlantis II Deep, Red Sea (21°18.58′N, 38°15.11′E). DSDP Leg 23, 225-26-CC (core-catcher). Drawn from a photomicrograph. Scale bar equals 0·05 mm.

Anhydrous calcium sulphate (anhydrite) may also be seen in evaporite sequences. Anhydrite can be extremely variable in form. Euhedral crystals tend to be prismatic or tabular and can resemble euhedral gypsum. However, grains may be denticulate, sub-spherical, bladed or rectangular (Fig. 55). Grains may show partial alteration to gypsum through hydration. Anhydrite shows cleavage in three directions (very perfect on {001} and {010} and less perfect on {100}), giving a cubic appearance. Like gypsum, anhydrite is colourless in plane-polarised light but has a higher relief which may show a slight change as the stage is rotated ($n_\alpha = 1{\cdot}570$, $n_\beta = 1{\cdot}576$ and $n_\gamma = 1{\cdot}614$). Extinction is parallel to the cleavage traces. In contrast to gypsum, anhydrite has high birefringence (0·044) and will therefore show bright interference colours.

HEAVY MINERALS

A wide range of detrital minerals may sometimes be observed in minor amounts in marine sediments. These include zircon, tourmaline, kyanite, sillimanite, rutile, anatase, epidote, garnet, andalusite, staurolite, sphene and many others. These minerals are generally uncommon in deep-sea sediments reflecting their low abundances in rocks and/or their relative instability in the marine environment. However, when they do occur, they may be exceptionally useful as clues to the nature of the source rocks and may be especially valuable in determining the source of turbidite beds. Many of these are termed 'heavy' minerals as they have specific gravities greater than 2·9 and hence sink in bromoform. This forming the basis of a separation technique which allows such dense minerals to be separated from the bulk of the sediment.

One method of separating heavy mineral suites from arenaceous sediments is to disaggregate samples by soaking in distilled water for 24 hours or as necessary. The samples are then dried, weighed and wet-sieved through a 63 μm sieve using distilled water, separating the sand fraction. The sand and mud (silt/clay) fractions are then dried and weighed. The sand is then separated into heavy and light fractions by centrifuge at 1500 rpm in tetrabromoform (specific gravity 2·89) for 15 minutes. Both fractions are then washed with acetone, dried and weighed. The heavy minerals are then separated into two size groups by sieving: >300 μm and 63–300 μm. Samples from the second group (63–300 μm) are mounted as grain mounts or smear slides and the different minerals identified by petrological examination (Holmes *et al.*, 1987). Although the presence of heavy minerals in smear slides should always be noted and if possible identified, specific studies of heavy mineral suites from marine sediments are best restricted to material coarser than 63 μm. Silt-sized heavy minerals, particularly if anhedral, may be difficult to identify specifically and may be too small for practical study. Carver (1971) and Allman & Lawrence (1972) provide a full discussion of separation techniques, while Hubert (1971) and Morton (1985) discuss the analysis of heavy mineral assemblages.

Heavy minerals may be divided into two groups: opaques and non-opaques. The non-opaque suite is the most useful in provenance studies and comprises a fairly homogeneous hydraulic suite, reducing

TABLE 8 Colour of Opaque Minerals in Reflected Light and Other Characteristics

Mineral	*Colour*	*Lustre*	*Grain shape*
Magnetite (Fe_3O_4)	Black, silver-grey, steel blue-black	Metallic	Octahedral, angular or well-rounded irregular grains
Ilmenite ($FeTiO_3$)	Blue-grey black	Purple-grey submetallic	Commonly occurs as irregular subangular grains
Hematite (Fe_2O_3)	Steel-grey black, reddish-brown	Metallic or earthy	Commonly occurs as irregular or rounded earthy grains
Limonite ($2Fe_2O_3.3H_2O$)	Dark brown, yellowish-brown	Submetallic, dull	Usually seen as irregular grains or powdery aggregates
Galena (PbS)	Lead grey	Metallic	Often shows cubic outlines containing right-angled steps
Pyrite (FeS_2)	Brass yellow	Metallic	Irregular masses, framboids, cubes and cube aggregates, fossil casts
Chalcopyrite ($CuFeS_2$)	Strong brass yellow	Metallic	Usually occurs as iridescent, irregular grains
Arsenopyrite (FeAsS)	Silver-white, steel grey	Metallic	May be crystalline or granular, very difficult to identify
Sphalerite (Zn, Fe)S	Black, brown, grey	Adamantine, metallic	Tetrahedral or dodecahedral grains or translucent brown granules
Gold (Au)	Golden yellow	Metallic	Rounded flattened grains of characteristic form.
Platinum (Pt)	White, aluminium grey	Metallic or dull	Usually seen as worn, rough flattened polygonal plates or irregular grains
Graphite (C)	Steel grey to soot black	Metallic or dull	Usually rounded, platy or irregular grains. Carbonaceous matter, although not strictly graphite, is often of similar appearance and is more common in sediments

problems of selective sorting. In heavy mineral analysis, opaque grains (such as ilmenite, pyrite, magnetite, marcasite) are often not identified according to mineral species, but grouped together. Non-opaque heavy minerals can be divided into two groups: the ultrastables—zircon, rutile and tourmaline—which are the most resistant to physical and chemical degradation (Pettijohn, 1941); and the metastables which include epidote, monazite, garnet, kyanite, sillimanite, sphene and staurolite. The combined percentage of the ultrastables (zircon, tourmaline and rutile) is termed the ZTR index (Hubert, 1962) and provides an index of mineralogical maturity.

The chemical stability of heavy minerals varies greatly and downhole the more unstable may show etched or corroded grain surfaces and skeletal grains (Edelman & Doeglas, 1932). However, not all ragged, 'hacksaw' grains are necessarily due to interstratal solution. It may be that some grain shapes merely reflect their original shape in the source rocks or breakage during transport.

Thirteen of the most common, easily identified, source-indicative or economically important heavy minerals are discussed in the following pages. All are non-opaque. Opaque minerals are not discussed in detail in this section although opaque iron oxides are discussed under the 'Iron Oxide' section of this book (pp. 139–143). Opaques usually only occur in very small amounts in most marine sediments (generally under 1%) although locally they can be important, especially in shelf sands or alluvial deposits derived from ore-bearing rocks. Opaques are best identified using reflected light and criteria such as lustre, colour, grain shape and surface features (Table 8). The shape of opaque grains may give clues to their nature. For instance, magnetite sometimes shows an octahedral form while galena grains often show cubic outlines containing right-angled steps. However, the study of opaques is a specialist study and often unequivocal identification requires specialist chemical techniques.

For workers engaged in local studies involving heavy mineral analysis, the best guide to optical identification still remains H. B. Milner's excellent text *Sedimentary Petrography* (Milner, 1962).

Andalusite

Detrital
Aluminium silicate
Al_2SiO_5

Form/Shape

Usually occurs as prismatic subhedral or angular anhedral grains, although grains can be very variable in form. Inclusions of quartz, biotite, iron oxides and/or carbonaceous matter may be present.

Transparency/Colour in PPL

Transparent to translucent, grains will be colourless or shades of pink (andalusite is pleochroic).

Relief

Moderate. Refractive indices: $n_\alpha = 1{\cdot}629\text{–}1{\cdot}640$, $n_\beta = 1{\cdot}633\text{–}1{\cdot}644$, $n_\gamma = 1{\cdot}639\text{–}1{\cdot}647$.

Birefringence

Weak (about 0·009), therefore grains will show low order interference colours in cross-polars.

Crystal System

Orthorhombic.

Cleavage

Prismatic cleavage, parallel to {110}, which may be conspicuous.

Pleochroism

In thin section, andalusite is weakly pleochroic from very pale pink to colourless. However in smear slides, grains may be thicker than normal thin sections and the pleochroism may be much more pronounced and may range to a rich pink colour. This is quite distinctive and almost diagnostic in itself.

Extinction

Prismatic grains will show straight extinction.

Optical Character

Biaxial, optically negative.

Occurrence/Source

Andalusite indicates a high rank metamorphic source. It has a moderate persistance and durability although Milner (1962) reports that as a detrital mineral, andalusite becomes increasingly rare with increasing geological age.

Main Distinguishing Features

Moderate relief, weak birefringence, straight extinction, prismatic cleavage, presence of inclusions and colourless–pink pleochroism.

Apatite

Detrital/authigenic
Calcium fluoro-phosphate
$Ca_5(PO_4)_3F$

Form/Shape

Usually occurs as euhedral or rounded subhedral six-sided prismatic crystals (Figs 56 and 57). Euhedral grains may show well-developed dome and prism faces. Well-rounded grains are often 'egg-shaped'.

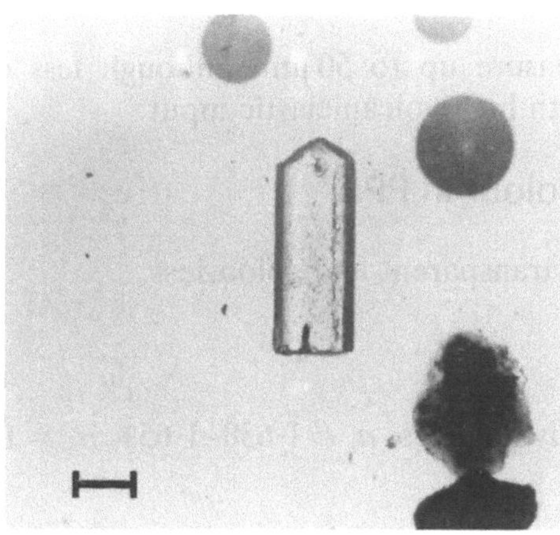

Fig. 56. Prismatic apatite grain. The two dark spherical objects in the upper part of the picture are air bubbles. Lower Cretaceous, lower continental rise near western edge of the Hatteras Abyssal Plain, Northwest Atlantic (35°29.71′N, 70°01.71′W). DSDP Leg 93, 603B-50-1, 113–114 cm. Plane-polarised light, scale bar equals 0·05 mm. (Courtesy of Drs M. A. Holmes, J. Breza, S. W. Wise and the Deep Sea Drilling Project.)

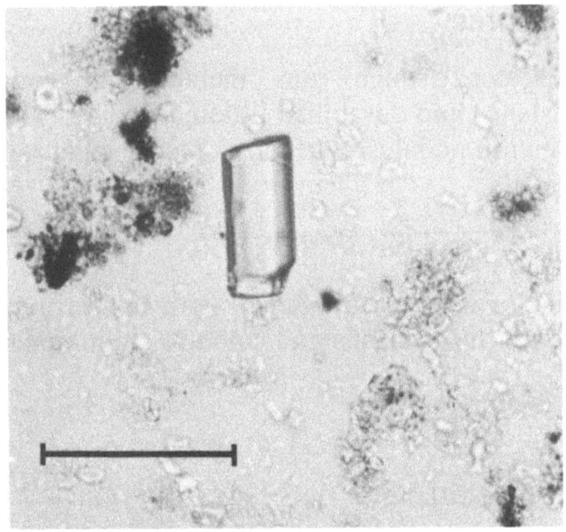

Fig. 57. Broken apatite prism. Maastrichtian, central Venezuelan Basin, Caribbean Sea (15°06.99′N, 69°22.67′W). DSDP Leg 15, 146-16-1, 122 cm. Plane-polarised light, scale bar equals 0·05 mm.

Grain Size Range

Grains usually measure up to 50 μm, although less commonly up to 500 μm in areas with high volcaniclastic input.

Transparency/Colour in PPL

Grains are usually transparent and colourless.

Relief

Moderate. Refractive indices: $n_\varepsilon = 1{\cdot}630–1{\cdot}651$, $n_\omega = 1{\cdot}633–1{\cdot}655$.

Birefringence

Weak (0·003–0·004), therefore grains will tend to show low order greys and whites in cross-polars.

Crystal System

Hexagonal.

Cleavage

Imperfect basal cleavage {0001} which may show as cross fractures.

Pleochroism

Absent.

Extinction

Straight.

Optical Character

Uniaxial, optically negative.

Occurrence/Source

Apatite is generally held to indicate a felsic igneous source, but it can form authigenically in areas where phosphate diagenesis is occurring (see Kolodny, 1981).

Main Distinguishing Features

Prismatic form, moderately high relief, weak birefringence, straight extinction and absence of colour.

Cassiterite

Detrital
Tin oxide
SnO_2

Form/Shape

Occurs as prismatic or pyramidal euhedra or subhedra or as rolled irregular anhedral grains. Prismatic grains may show fluted or finely striated faces, the striations being parallel to the principal axis.

Transparency/Colour in PPL

May be translucent or opaque. Translucent grains are yellowish, reddish or brown in colour and coloration is often uneven. Translucent grains often contain inclusions, usually of iron oxides, and may show zoning. Opaque grains may show translucent areas.

Relief

Very high. Refractive indices: $n_\omega = 1{\cdot}966$, $n_\varepsilon = 2{\cdot}093$. Larger grains may show an adamantine lustre in reflected light.

Birefringence

Very high (0·097). However, the high order polarisation colours are usually masked by the mineral's natural colour.

Crystal System

Tetragonal.

Cleavage

Imperfect {100}, poor {111}.

Pleochroism

Cassiterite may occasionally be pleochroic in browns, yellows or reds. Milner (1962) reports that pleochroism is sometimes consistently present in samples from particular localities.

Extinction

Prismatic grains show straight extinction.

Optical character

Uniaxial, optically positive.

Occurrence

Detrital cassiterite will be derived from veins, granite pegmatites and metalliferous lodes. Cassiterite is an important ore of tin and economic placers may be located offshore of existing onshore stream placers, particularly when these pass into drowned river valleys (Emery & Noakes, 1968). Major cassiterite placers occur off Thailand, Malaysia and Indonesia and some of these are described by Overeem (1960) and Aleva (1973). These are alluvial placers that have been submerged during the postglacial rise in sealevel (Emery & Noakes, 1968). Similar deposits have been reported off Cornwall where drilling has revealed the presence of cassiterite near bedrock and within the sediment column in several drowned river valleys (Cronan, 1980).

Main Distinguishing Features

Form, colour, relief, extinction and birefringence. However, cassiterite is a difficult mineral to positively identify optically. It may be confused with other dusky brown minerals such as sphene, rutile, sphalerite or hypersthene. It may be distinguished from rutile by its lower birefringence while sphalerite is isotropic.

Epidote

Detrital
Hydrous calcium aluminium iron silicate
$Ca_2(Al, Fe)_3(OH)(SiO_4)_3$

Form/Shape

Usually occurs as prismatic subhedra (Fig. 58) or elongate anhedra (Fig. 59).

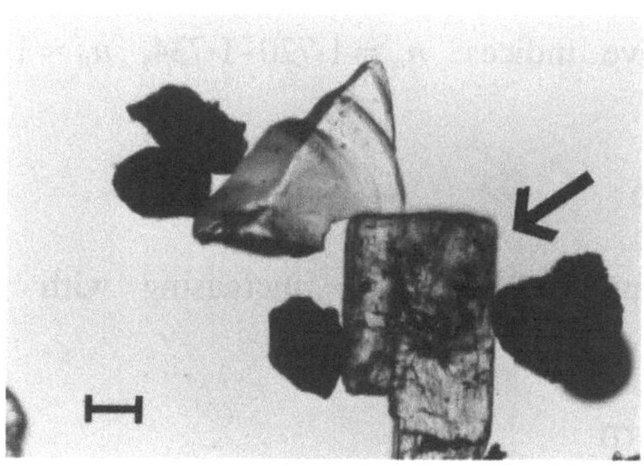

Fig. 58. Prismatic epidote grain (arrowed). The irregular grain to the upper left is garnet. Lower Cretaceous, lower continental rise near western edge of Hatteras Abyssal Plain, Northwest Atlantic (35°29.71′N, 70°01.71′W). DSDP Leg 93, 603B-57-5, 37–38 cm. Plane-polarised light, scale bar equals 0·05 mm. (Courtesy of Drs M. A. Holmes, J. Breza, S. W. Wise and the Deep Sea Drilling Project.)

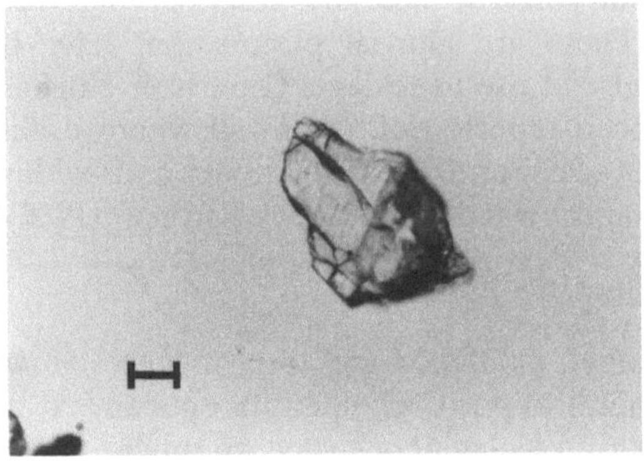

Fig. 59. Elongate anhedral epidote grain. Lower Cretaceous, lower continental rise near western edge of Hatteras Abyssal Plain, Northwest Atlantic (35°29.71′N, 70°01.71′W). DSDP Leg 93, 603B-57-5, 37–38 cm. Plane-polarised light, scale bar equals 0·05 mm. (Courtesy of Drs M. A. Holmes, J. Breza, S. W. Wise and the Deep Sea Drilling Project.)

Transparency/Colour in PPL

Usually colourless and transparent. Iron-rich varieties may be greenish-yellow, although colour may not be uniform.

Relief

High. Refractive indices: $n_\alpha = 1{\cdot}720\text{–}1{\cdot}734$, $n_\beta = 1{\cdot}724\text{–}1{\cdot}763$, $n_\gamma = 1{\cdot}734\text{–}1{\cdot}779$.

Birefringence

Moderate–strong (0·014–0·045), increasing with increase in iron content.

Crystal System

Monoclinic.

Cleavage

Perfect {001}, imperfect {100}.

Pleochroism

Grains may be slightly pleochroic in yellowish greens. Pleochroism is usually more noticeable in thick grains.

Extinction

Straight.

Optical Character

Biaxial, optically negative.

Occurrence/Source

Detrital epidote is usually indicative of erosion of regionally metamorphic terrains.

Main Distinguishing Features

Yellowish-green colour and weak pleochroism in PPL (iron-rich varieties), high relief, moderate–strong birefringence and straight extinction in prismatic grains.

Garnet

Detrital

The commoner members of the garnet group are aluminium silicates of magnesium, iron, manganese or calcium.

General formula: $A_3B_2Si_3O_{12}$ where A can be Fe^{2+}, Mg, Mn or Ca and B can be Fe^{3+}, Al or Cr.

The commoner members of the group are:

Almandine	$Fe_3Al_2Si_3O_{12}$
Pyrope	$Mg_3Al_2Si_3O_{12}$
Spessartite	$Mn_3Al_2Si_3O_{12}$
Grossularite	$Ca_3Al_2Si_3O_{12}$
Andradite	$Ca_3Fe_2Si_3O_{12}$

Form/Shape

Usually occurs as small almost circular grains which may be densely packed with inclusions of other minerals. However, grains can be quite

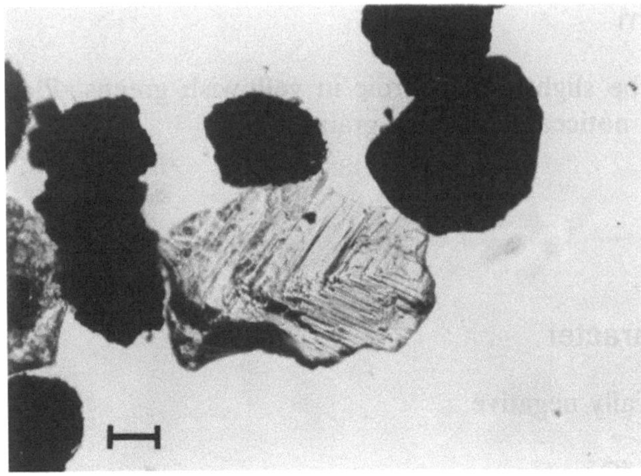

Fig. 60. Subangular garnet grain showing extensive chemical etching, resulting in steplike V-shaped surface features. Lower Cretaceous, lower continental rise near western edge of Hatteras Abyssal plain, Northwest Atlantic (35°29.71′N, 70°01.71′W). DSDP Leg 93, 603B-58-1, 50–51 cm. Plane-polarised light, scale bar equals 0·05 mm. (Courtesy of Drs M. A. Holmes, J. Breza, S. W. Wise and the Deep Sea Drilling Project.)

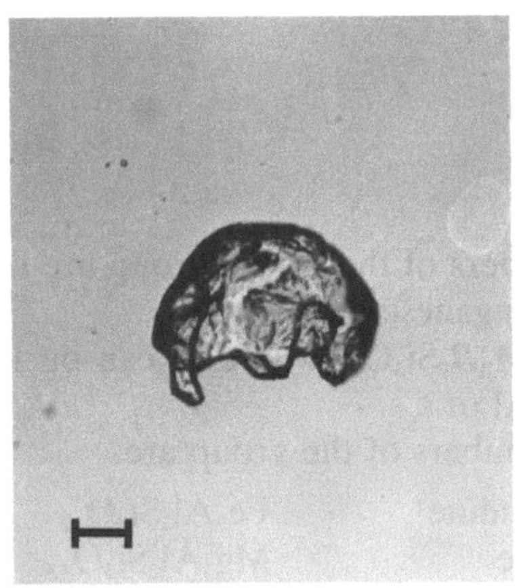

Fig. 61. Anhedral subrounded garnet grain showing extensive chemical etching that has resulted in embayed surface features. Lower Cretaceous, lower continental rise near western edge of Hatteras Abyssal Plain, Northwest Atlantic (35°29.71′N, 70°01.71′W). DSDP Leg 93, 603B-71-1, 116–117 cm. Plane-polarised light, scale bar equals 0·05 mm. (Courtesy of Drs M. A. Holmes, J. Breza, S. W. Wise and the Deep Sea Drilling Project.)

irregular. Grains are often grooved, pitted or marked by rectangular patterning or other surface features through chemical etching (Figs 60 and 61). Grains which have had a long residence time in the sedimentary cycle or experienced high environmental energies may show irregular fracture surfaces.

Transparency/Colour in PPL

Transparent–translucent, grains may vary from colourless to pale pinks or browns. Grossular garnet may be yellow, brown or green.

Relief

Very high. Refractive index varies with composition, $n = 1{\cdot}74\text{–}1{\cdot}89$. Garnets will therefore tend to be prominent in the field of view.

Birefringence

Being isotropic, garnets will appear dark between cross-polars, although strain, crystal zoning or twinning may cause sectors of weak birefringence. Inclusions within garnet grains are frequently anisotropic.

Crystal System

Cubic.

Cleavage

Absent, although complex irregular fractures are typically developed.

Pleochroism

Absent.

Occurrence/Source

Garnets indicate a high rank metamorphic source. The type of garnet present in a sediment may be identified using electron microprobe analysis. Grossular garnet is especially characteristic of thermally and regionally metamorphosed impure calcareous rocks, although most garnet seen in marine sediments is detrital. Donnelly & Nalli (1973)

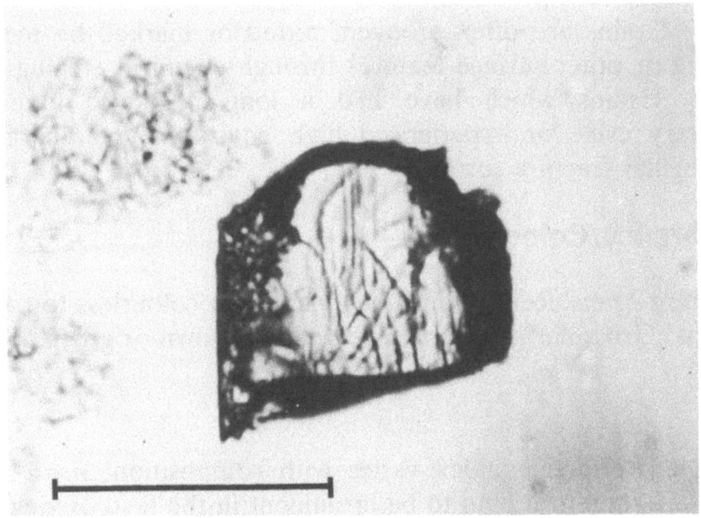

Fig. 62. Spessartine garnet. Maastrichtian, central Venezuelan Basin, Caribbean Sea (15°06.99′N, 69°22.67′W). DSDP Leg 15, 146-15-4, 10 cm. Plane-polarised light, scale bar equals 0·05 mm. (Courtesy of Drs T. W. Donnelly and G. Nalli and the Deep Sea Drilling Project.)

suggest an authigenic origin for spessartine garnet occurring in Maastrichtian marl ooze from the Venezuela Basin (DSDP Hole 146) (Fig. 62).

Main Distinguishing Features

Isotropism, high refractive index, form, absence of cleavage and colour.

Kyanite

Detrital
Aluminium silicate
Al_2SiO_5

Form/Shape

Typically occurs as bladed flakes with irregular terminations or sharply angular cleavage fragments. Bent and distorted flake are not uncommon. Grains may appear ‘frayed’ due to corrosion.

Transparency/Colour in PPL

Flakes will be transparent and vary from colourless to a pale blue. Blue coloration may be patchy.

Relief

High. Refractive indices: $n_\alpha = 1{\cdot}712$, $n_\beta = 1{\cdot}720$, $n_\gamma = 1{\cdot}728$.

Birefringence

Moderate (0·016).

Crystal System

Triclinic.

Cleavage

Kyanite has three cleavage directions, perfect parallel to {100}, less perfect parallel to {010} and a more widely spaced parting, rather than a true cleavage, parallel to {001}.

Extinction

Kyanite shows oblique extinction to the cleavage traces up to 30° on {100}.

Optical Character

Biaxial, optically negative.

Occurrence/Source

Kyanite indicates a high rank metamorphic source and will frequently be associated with other detrital metamorphic minerals such as staurolite, garnet and andalusite.

Main Distinguishing Features

Colour, grain shape, high relief and conspicuous cleavage. The bladed grain shape, controlled by the cleavage, is distinctive, making kyanite an easy mineral to identify.

Monazite

Detrital
A rare earth phosphate
(Ce, La, Nd, Pr)PO_4

Form/Shape

Monazite typically occurs as small prismatic euhedral or subhedral crystals or rounded egg-shaped grains.

Transparency/Colour in PPL

Transparent and colourless, although thick grains may be pale yellow or even brownish in colour.

Relief

Very high. Refractive indices: $n_\alpha = 1{\cdot}786–1{\cdot}800$, $n_\beta = 1{\cdot}788–1{\cdot}801$, $n_\gamma = 1{\cdot}837–1{\cdot}849$.

Birefringence

Strong to very strong (0·049–0·051). Grains will therefore show high order interference colours in cross-polars.

Crystal system

Monoclinic.

Cleavage

Perfect {001}, good {100}.

Pleochroism

Usually absent, although thick grains may be pleochroic in yellows.

Extinction

Prismatic grains will show a small extinction angle of 2–10° with respect to the *c* axis.

Optical Character

Biaxial, optically positive.

Occurrence/Source

Monazite indicates a felsic igneous source and may be a relatively common mineral in sands eroded from granitic hinterlands. It is a persistent mineral and is resistant to intrastratal solution.

Main Distinguishing Features

Monazite may be difficult to identify with certainty in smear slides but form, the typically small grain size and the very high relief are probably the most diagnostic criteria. Monazite closely resembles sphene in optical properties but has a lower birefringence.

Rutile

Detrital
Titanium dioxide
TiO_2

Form/Shape

Rutile occurs as small prismatic, often slightly rounded, grains or rare geniculate twins. Euhedral crystals have pyramidal terminations. The degree of rounding is often less than associated minerals. Many prismatic grains show characteristic striations running obliquely to prism edges due to polysynthetic twinning.

Transparency/Colour in PPL

Transparent and usually yellow to reddish-brown in colour.

Relief

Exceptionally high. Refractive indices: $n_{\omega} = 2{\cdot}603–2{\cdot}616$, $n_{\varepsilon} = 2{\cdot}889–2{\cdot}903$. Grain outlines will appear especially prominent due to the high refractive index.

Birefringence

Extreme (0·286–0·287), however, the very high order interference colours are usually obscured by total reflection.

Crystal System

Tetragonal.

Cleavage

Parallel to the length of the crystals {110}.

Pleochroism

Reddish-brown varieties are sometimes weakly pleochroic with the maximum absorption direction parallel to the principal axis.

Extinction

Straight.

Optical Character

Uniaxial, optically positive.

Occurrence/Souce

Rutile is sometimes considered to indicate a basic igneous source (Pettijohn, 1975), although it is a rather widely distributed accessory mineral in some igneous and metamorphic rocks. Like zircon, it is highly durable and is often a common component of heavy mineral suites separated from arenaceous sediments.

Main Distinguishing Features

Prismatic form, colour, cleavage, extreme birefringence, the presence of diagonal striae when developed and exceptionally high relief.

Sillimanite

Detrital
Aluminium silicate
Al_2SiO_5

Form/Shape

Usually occurs as small, slender, prismatic grains, often with irregular or fractured terminations.

Transparency/Colour in PPL

Transparent and colourless, although thick grains may appear yellow or brownish.

Relief

High. Refractive indices: $n_\alpha = 1{\cdot}657–1{\cdot}661$, $n_\beta = 1{\cdot}658–1{\cdot}670$, $n_\gamma = 1{\cdot}677–1{\cdot}684$.

Birefringence

Moderate (about 0·021).

Crystal System

Orthorhombic.

Cleavage

Parallel to {010}, although cleavage is not normally conspicuous. Transverse fractures are common.

Pleochroism

Absent.

Extinction

Extinction is parallel to the prism edges.

Optical Character

Biaxial, optically positive.

Occurrence/Source

Sillimanite indicates a high rank metamorphic source.

Main Distinguishing Features

High relief, moderate birefringence, straight extinction and grain shape. Sillimanite is rather undistinctive and may be confused with andalusite, although it has a stronger birefringence. In plane-polarised light sillimanite can sometimes resemble kyanite, but may be distinguished by its straight extinction, lower refractive index and higher interference colours.

Sphene (Titanite)

Detrital
Calcium titanium silicate
$CaTiSiO_5$

Form/Shape

Sphene occurs as small diamond-shaped eu/subhedra or irregular, sometimes acutely ragged, grains. Grains frequently contain inclusions.

Transparency/Colour in PPL

Grains will be colourless to brown or brownish-yellow. Orange and greenish varieties do occur but these are rare.

Relief

Very high. Refractive indices: $n_\alpha = 1{\cdot}887–1{\cdot}913$, $n_\beta = 1{\cdot}894–1{\cdot}921$, $n_\gamma = 1{\cdot}979–2{\cdot}054$.

Birefringence

Very strong (0·092–0·141), therefore interference colours will be high-order but are usually obscured by total reflection.

Crystal System

Monoclinic.

Cleavage

{110}, distinct. Sphene also often shows a prominent parting parallel to {221}.

Pleochroism

Some strongly coloured varieties are pleochroic in yellows and browns.

Extinction

Grains will rarely show complete extinction due to strong dispersion.

Optical Character

Biaxial, optically positive. Sphene has a very strong axial dispersion, giving a rather peculiar interference figure with convergent light. Although usually incomplete in detrital grains, the interference figure may have some diagnostic value.

Occurrence

Like monazite, sphene usually indicates a felsic igneous source, particularly when it occurs as euhedra or subhedra. However, it also occurs in certain metamorphic rocks, particularly hornblende schists and amphibolites.

Main Distinguishing Features

Characteristic wedge-shaped crystals when euhedral, colour, slight pleochroism, well-developed cleavage traces not parallel to bounding edges (when euhedral or subhedral), high relief, incomplete extinction and strong birefringence. However, irregular grains may still prove difficult to positively identify.

Staurolite

Detrital
A complex hydrous iron aluminium silicate
$2Al_2SiO_5.Fe(OH)_2$

Form/Shape

Staurolite usually occurs as irregular, anhedral grains (Figs 63 and 64), which may show serrated edges, reflecting its prismatic cleavage. It may also occur as prismatic suhedra or more rarely, euhedra. Grains commonly contain inclusions. Grain surfaces may be etched and pitted (Fig. 64).

Transparency/Colour in PPL

Grains are transparent and usually pale yellow to a medium yellowish-brown. Thick grains may appear translucent.

Relief

High, refractive indices: $n_\alpha = 1{\cdot}736–1{\cdot}747$, $n_\beta = 1{\cdot}741–1{\cdot}754$, $n_\gamma = 1{\cdot}746–1{\cdot}762$.

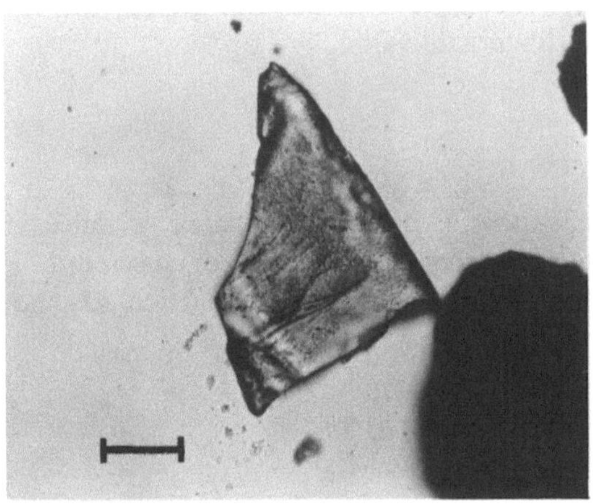

Fig. 63. Angular staurolite grain. Lower Cretaceous, lower continental rise near the edge of the Hatteras Abyssal Plain, Northwest Atlantic (35°29.71′N, 70°01.71′W). DSDP Leg 93, 603B-52-5, 106–107 cm. Plane-polarised light, scale bar equals 0·05 mm. (Courtesy of Drs M. A. Holmes, J. Breza, S. W. Wise and the Deep Sea Drilling Project.)

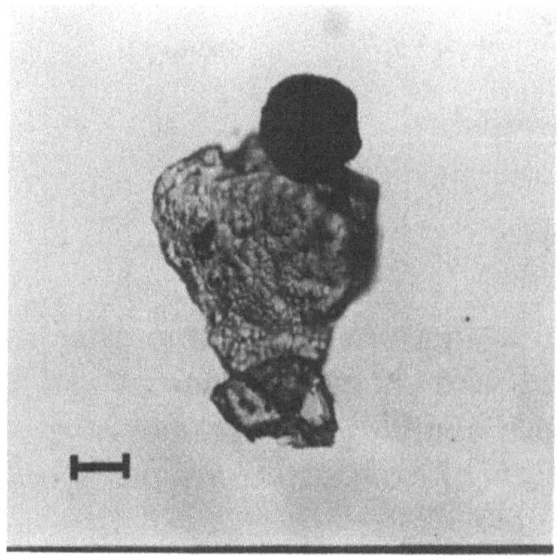

Fig. 64. Anhedral staurolite grain showing hummocky surface features caused by chemical etching. Lower Cretaceous, lower continental rise near western edge of Hatteras Abyssal Plain, Northwest Atlantic (35°29.71′N, 70°01.71′W). DSDP Leg 93, 603B-63-7, 16–17 cm. Plane-polarised light, scale bar equals 0·05 mm. (Courtesy of Drs M. A. Holmes, J. Breza, S. W. Wise and the Deep Sea Drilling Project.)

Birefringence

Weak (0·010–0·015), therefore grains will tend to show low order interference colours between cross-polars.

Crystal System

Orthorhombic.

Cleavage

Prismatic cleavage, parallel to {010}. Cleavage is usually not conspicuous but may affect grain fragmentation, giving grains ‘saw-tooth’ edges.

Pleochroism

Staurolite is distinctly pleochroic from pale yellows (almost colourless in small grains) to medium yellows/browns, although its intensity varies greatly.

Extinction

Prismatic sections show straight extinction.

Optical Character

Biaxial, optically positive.

Occurrence

Staurolite indicates a high rank metamorphic source. Staurolite is commonly associated with sillimanite, garnet and kyanite in metamorphic rocks and these tend to be commonly associated minerals in detrital sediments derived from such terrains.

Main Distinguishing Features

Colour, high relief, pleochroism, straight extinction, association with other detrital high rank metamorphic minerals such as kyanite.

Tourmaline

Detrital
Complex aluminium borosilicate

Form/Shape

Occurs as prismatic eu/subhedral crystals (Figs 65 and 66) and fractured irregular to well-rounded grains. Prismatic grains are often terminated by basal parting and therefore may lack terminal faces. Grains often contain inclusions.

Transparency/Colour in PPL

Transparent but can show a great variety of colour including shades of blue, green, pink, brown and ranging to a pale buff or even colourless in the position of minimum absorption (tourmaline is pleochroic).

Relief

Moderate to high. Refractive index varies with composition. Range of refractive indices: $n_\varepsilon = 1{\cdot}613\text{–}1{\cdot}658$, $n_\omega = 1{\cdot}632\text{–}1{\cdot}698$, depending on variety.

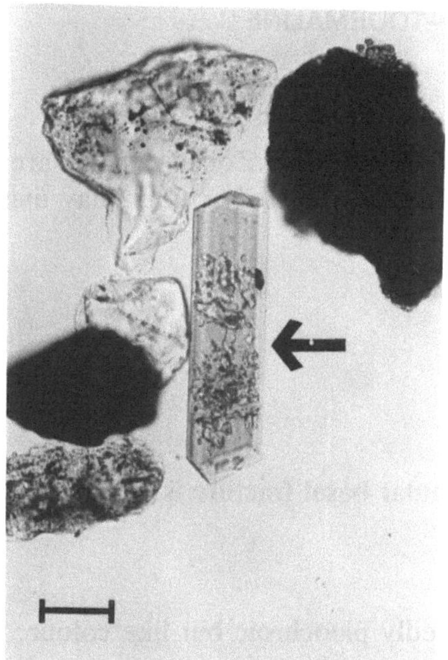

Fig. 65. Prismatic, euhedral tourmaline crystal (arrowed) showing etched surface features. Lower Cretaceous, lower continental rise near western edge of Hatteras Abyssal Plain, Northwest Atlantic (35°29.71′N, 70°01.71′W). DSDP Leg 93, 603B-63-7, 16–17 cm. Plane-polarised light, scale bar equals 0·05 mm. (Courtesy of Drs M. A. Holmes, J. Breza, S. W. Wise and the Deep Sea Drilling Project.)

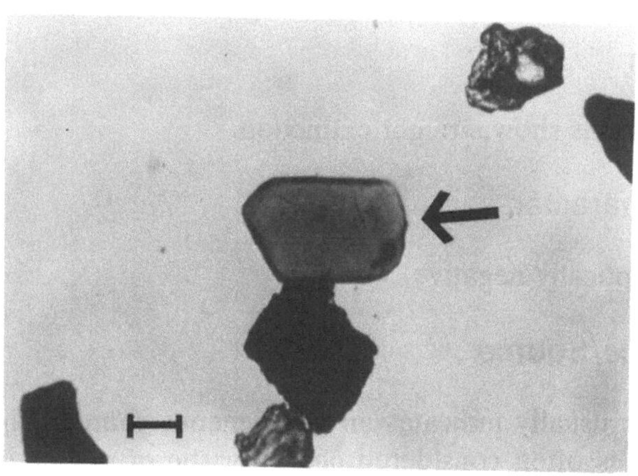

Fig. 66. Subrounded euhedral tourmaline grain. Lower Cretaceous, lower continental rise near western edge of Hatteras Abyssal Plain, Northwest Atlantic (35°29.71′N, 70°01.71′W). DSDP Leg 93, 603B-63-7, 16–17 cm. Plane-polarised light, scale bar equals 0·05 mm. (Courtesy of Drs M. A. Holmes, J. Breza, S. W. Wise and the Deep Sea Drilling Project.)

Birefringence

Moderate (usually 0·015–0·025). Zoned crystals are quite common. In black tourmaline (schorl), the body colour may mask the interference colours.

Crystal System

Hexagonal.

Cleavage

Absent, but an irregular basal fracture is commonly developed.

Pleochroism

Tourmaline is markedly pleochroic but like colour, pleochroism varies greatly. Brown and dark coloured grains are usually the most strongly pleochroic. The blue variety, indicolite, is usually only weakly pleochroic. Prisms and grains are darkest when the *c* axis (this direction can be easily identified, even in anhedra, as it will be normal to the direction of the basal {0001} trending fractures) is at right-angles to the vibration direction of the polariser. In thick detrital grains, absorption may be complete so that in one stage position, the grain will appear quite black.

Extinction

Prismatic grains show straight extinction.

Optical Character

Uniaxial, optically negative.

Occurrence/Source

Tourmaline usually indicates an acid igneous source. The blue variety (indicolite) is often considered characteristic of pegmatites (Pettijohn, 1975). Tourmaline is also found in some metamorphic rocks, for example the brown variety, dravite, usually occurs in metamorphosed limestones. Tourmaline having a metamorphic source may contain numerous inclusions. Tourmaline has a high durability and a high persistence. Rounded tourmaline usually indicates reworked sediments.

Main Distinguishing Features

Prismatic form when subhedral or euhedral, presence of basal {0001} trending fractures, straight extinction and marked pleochroism.

Zircon

Detrital
Zirconium silicate
$ZrSiO_4$

Form/Shape

Zircon occurs as elongate to almost equant euhedral or subhedral crystals (Figs 67–70) and more rarely, rounded anhedral grains. Euhedral crystals will show prismatic form with well-developed dome and pyramid faces (Fig 67). Zircon grains frequently contain inclusions (Figs 69 and 70) which are usually irregularly distributed within the grain but may sometimes be aligned parallel to the length of the crystal.

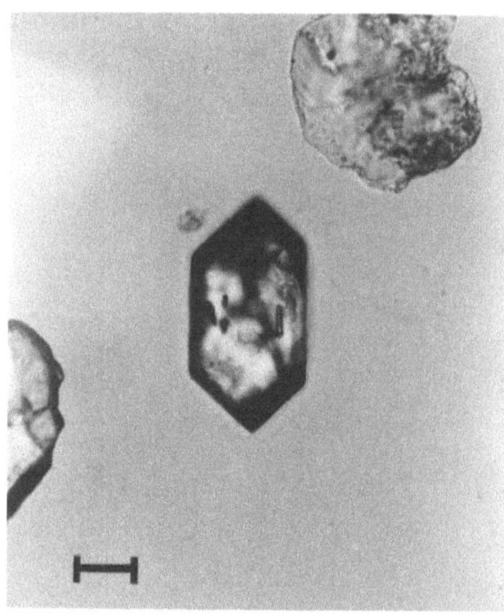

Fig. 67. Euhedral zircon crystal containing tubular inclusions. Note the dark outline of the crystal due to the very high relief. Lower Cretaceous, lower continental rise near western edge of Hatteras Abyssal Plain, Northwest Atlantic (35°29.71′N, 70°01.71′W). DSDP Leg 93, 603B-63-7, 16–17 cm. Plane-polarised light, scale bar equals 0·05 mm. (Courtesy of Drs M. A. Holmes, J. Breza, S. W. Wise and the Deep Sea Drilling Project.)

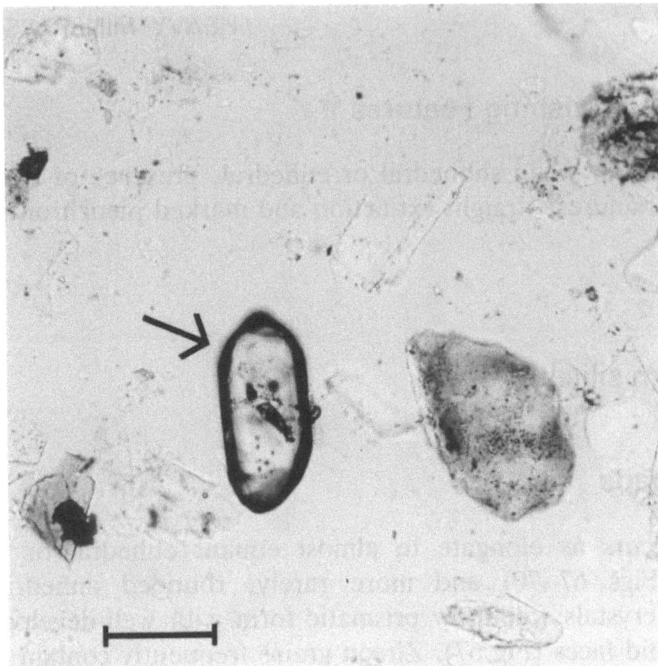

Fig. 68. Prismatic subeuhedral zircon crystal (arrowed). Note the very high relief which makes the grain prominent in the field of view. Pliocene, south of Java Trench, Northeast Indian ocean (9°46.53′S, 102°41.95′E). DSDP Leg 22, 211-8-1, 63 cm. Plane-polarised light, scale bar equals 0·05 mm.

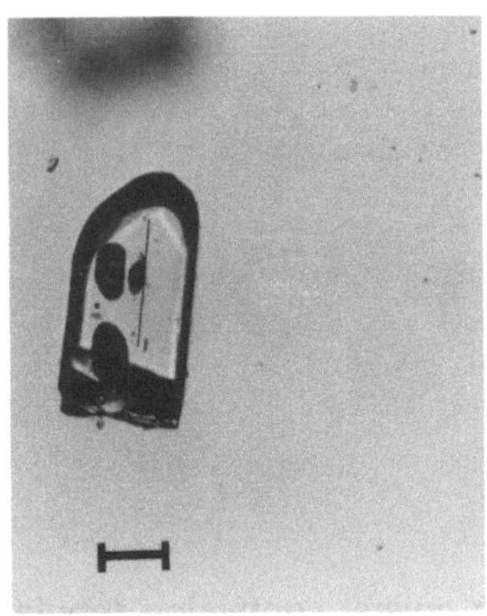

Fig. 69. Prismatic zircon grain containing inclusions. Lower Cretaceous, lower continental rise near western edge of Hatteras Abyssal Plain, Northwest Atlantic (35°29.71′N, 70°01.71′W). DSDP Leg 93, 603B-58-1, 50–51 cm. Plane-polarised light, scale bar equals 0·05 mm. (Courtesy of Drs M. A. Holmes, J. Breza, S. W. Wise and the Deep Sea Drilling Project.)

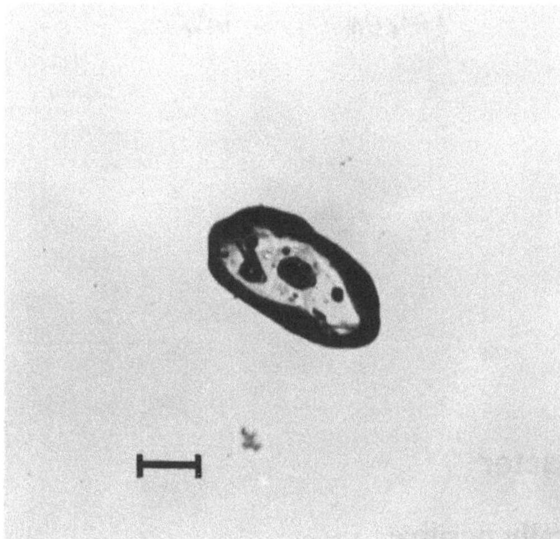

Fig. 70. Anhedral subrounded zircon grain containing numerous round inclusions. Lower Cretaceous, lower continental rise near western edge of Hatteras Abyssal Plain, Northwest Atlantic (35°29.71′N, 70°01.71′W). DSDP Leg 93, 603B-58-1, 50–51 cm. Plane-polarised light, scale bar equals 0·05 mm. (Courtesy of Drs M. A. Holmes, J. Breza, S. W. Wise and the Deep Sea Drilling Project.)

Transparency/Colour in PPL

Usually colourless and transparent or more rarely faint shades of yellow or brown. Faint pink, mauve and even green examples have been described but these are rare.

Relief

Very high, consequently zircon crystals will be very prominent in the field of view. Refractive indices: $n_\omega = 1{\cdot}925\text{–}1{\cdot}931$, $n_\varepsilon = 1{\cdot}985\text{–}1{\cdot}993$.

Birefringence

Very strong (0·060–0·062) therefore interference colours will be high order. In anhedral grains, colour fringes of lower interference colours may be seen towards the grain margins.

Crystal System

Tetragonal.

Cleavage

Absent.

Pleochroism

Absent.

Extinction

Straight.

Optical Character

Uniaxial, optically positive.

Occurrence/Source

Euhedral zircon usually indicates a felsic igneous source, although it does occur in certain metamorphic rocks. It has a high persistence and rounded zircon usually indicates reworked sediments. It is a frequent component of heavy mineral suites separated from arenaceous sediments.

Main Distinguishing Features

Crystal form, transparency, presence of inclusions, lack of colour, straight extinction, strong birefringence and very high relief.

IRON OXIDES

Detrital/authigenic

Non-opaque iron oxides that occur in marine sediments include flakes of red haematite and grains of brown hydrated iron oxide (limonite). Opaque iron oxides include haematite, limonite, magnetite and ilmenite.

Haematite

Fe_2O_3

Form/Shape

Irregular flecks, flakes, scales (Fig. 71) and irregular to rounded earthy particles. More rarely, it may occur as rhombohedral or platy grains.

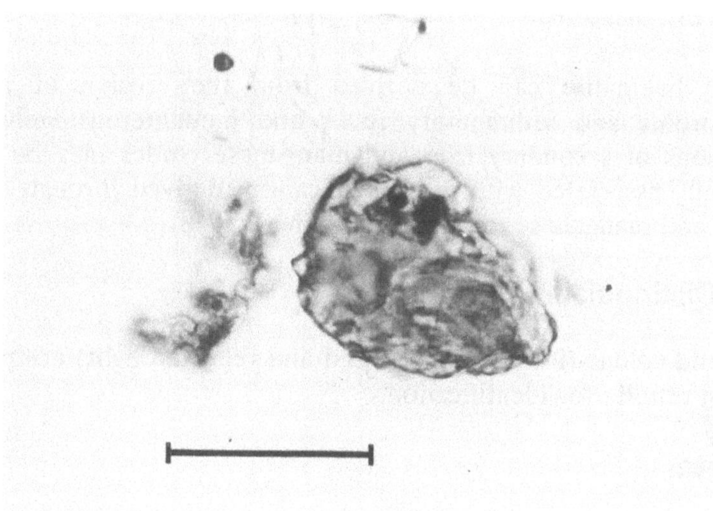

Fig. 71. Small, irregular, iron oxide scale, tentatively identified as haematite, reddish-brown in transmitted light. Upper Eocene, Kings Trough, Northeast Atlantic (44°02.12′N, 21°48.79′W). Plane-polarised light, scale bar equals 0·05 mm.

Grain Size Range

Very variable, covering the clay to sand range.

Transparency/Colour in PPL

Grains may be translucent or opaque. Translucent grains may be ruby red or any shade of brown. Colour and translucency may be uneven. Opaque grains are reddish-brown by reflected light.

Relief

Translucent grains show a high relief. Refractive indices: $n_\varepsilon = 2{\cdot}94$, $n_\omega = 3{\cdot}22$.

Birefringence

High (0·28), however, polarisation effects are usually masked by body colour.

Occurrence/Source

Detrital haematite can be derived from the erosion of igneous, metamorphic and sedimentary rocks and metalliferous veins. Concentrations of secondary iron and manganese oxides may be characteristically associated with marine volcanism derived through leaching of iron and manganese from lavas (Bonnati, 1968).

Main Distinguishing Features

Form and colour (in both transmitted and reflected light) are probably the best criteria for identification.

Ilmenite

$FeTiO_3$

Form/Shape

Commonly occurs as irregular, sub-angular grains.

Transparency/Colour in PPL

Grains are quite opaque in transmitted light. In reflected light ilmenite will appear a blue-grey black and show a metallic lustre.

Occurrence/Source

Ilmenite may be a relatively common opaque in sediments derived from igneous rocks (especially basic or ultrabasic types).

Ilmenite will often be difficult to distinguish from magnetite in smear slides. These two minerals are probably best classed together in initial smear slide descriptions.

Limonite

$2Fe_2O_3.3H_2O$ (?)

Form/Shape

Irregular grains (Fig. 72) and amorphous masses. Hydrated iron oxides commonly coat detrital particles.

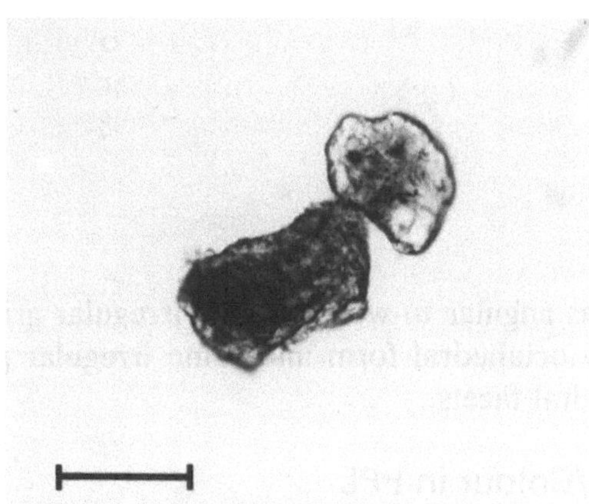

Fig. 72. Irregular limonite grains, brown in transmitted light. Pleistocene, Madeira Abyssal Plain, Northeast Atlantic. Plane-polarised light, scale bar equals 0·05 mm.

Transparency/Colour in PPL

Large grains are opaque but smaller grains may be translucent and yellowish-dark brown in colour. Opaque grains are brown in reflected light.

Birefringence

The strong body colour of translucent grains usually masks any polarisation effects.

Occurrence/Source

Limonite is a common secondary mineral formed primarily as a result of oxidation or weathering. The iron is usually derived from iron-bearing minerals or iron ores. Hydrated iron oxides may be common in submarine volcanic provinces (Bonatti, 1968). Leaching of iron and manganese from lavas would be enhanced through high temperatures and through the presence of volcanic gases dissolved in seawater during eruptions. According to Bonatti (1968) convection currents initiated by the eruption, passing through hyaloclastites, would help supply enough oxygen to precipitate the iron and manganese oxides. Other sources for the iron include pyrite and glauconite being displaced through mass gravity flows from shallower to deeper water. These minerals may be partially or completely oxidised yielding iron oxides. Conversely, terrigenous iron oxide may be the source of much of the iron in hemipelagic pyrites.

Magnetite

Fe_3O_4

Form/Shape

Usually occurs as angular to well-rounded irregular grains. Grains may sometimes show octahedral form and some irregular grains may show traces of octahedral facets.

Transparency/Colour in PPL

Magnetite is always opaque in transmitted light. In reflected light, it will appear grey-black and show a metallic, submetallic or even dull lustre.

Occurrence/Source

Magnetite is a common accessory mineral in most igneous and metamorphic rocks and may be a common detrital mineral in sediments derived from these. Octahedral grains are often distinctive but when irregular, grains may be very difficult to distinguish from ilmenite.

MICA

Detrital

The micas are a group of hydrous potassium aluminium silicates. The most common members of this group are muscovite, biotite and chlorite.

Muscovite $KAl_2(OH)_2(AlSi_3O_{10})$

Biotite $K_2(Mg, Fe)_2(OH)_2(AlSi_3O_{10})$

Chlorite is, in fact, a group of hydrous magnesium and iron aluminium silicates. Several species within the chlorite group are recognised and they frequently form as alteration products of other ferromagnesian minerals. In smear slide description, it is customary to use the word 'chlorite' to describe rather nondescript, fine-grained, green or greenish-yellow, blue-polarising grains and flakes that cannot be more precisely identified due to lack of definite optical properties. Milner (1962) recommends that such grains should be described as 'chloritic matter' rather than 'chlorite'.

Form/Shape

Usually seen as ragged flakes or scales which are often conspicuously larger than associated material (Figs 73–78). Flakes often show traces of successive {001} cleavage surfaces exposed through breakage (Fig. 73). Biotite may sometimes show a pseudohexagonal form. Grains may contain mineral inclusions.

Fragmentation Characteristics

Micas are generally resistant to fragmentation (a reflection of their flaky habit). However, shredding of flakes may produce irregular flecks.

Grain Size Range

General size range 10–1000 μm.

Transparency/Colour in PPL

Transparent, muscovite is colourless; biotite is green or brown and chloritic matter is usually varying shades of green or greenish-yellow.

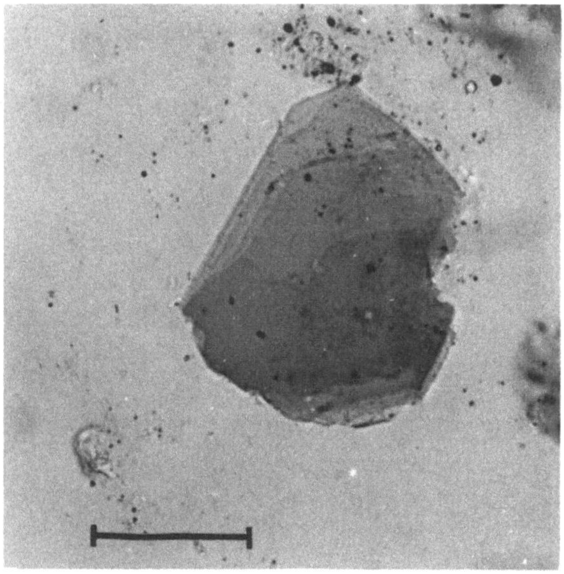

Fig. 73. Muscovite flake. Successive {001} cleavage surfaces exposed through breakage are clearly visible. Pleistocene, Tyrrhenian Sea, Mediterranean. Plane-polarised light, scale bar equals 0·05 mm.

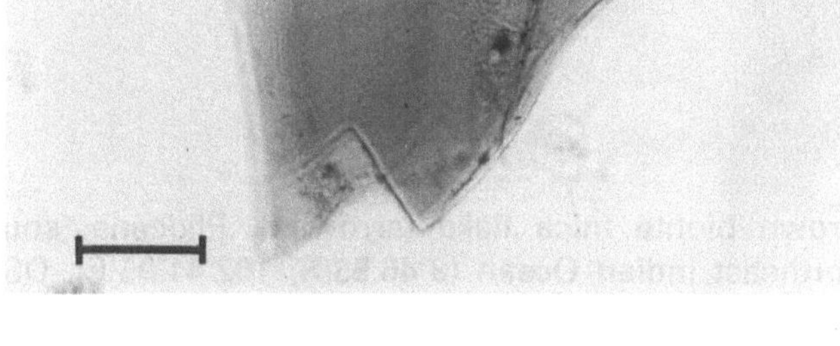

Fig. 74. Biotite mica, brown in transmitted light. Pleistocene, Tyrrhenian Sea, Mediterranean. Plane-polarised light, scale bar equals 0·05 mm.

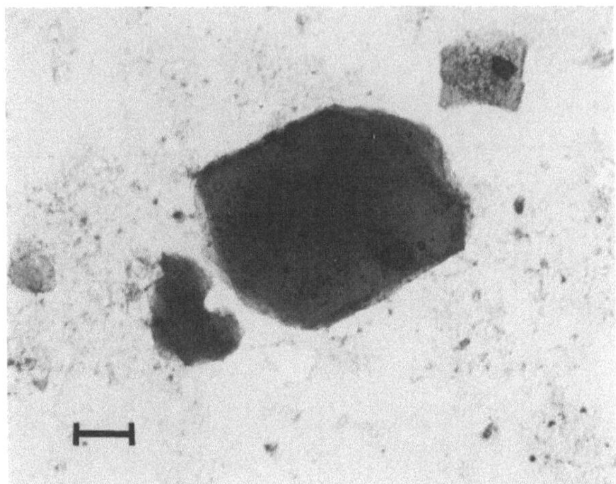

Fig. 75. Flakes of green biotite. Pliocene, south of Java Trench, Northeast Indian Ocean (9°46.53'S, 102°41.95'E). DSDP Leg 22, 211-8-1, 56 cm. Plane-polarised light, scale bar equals 0·05 mm.

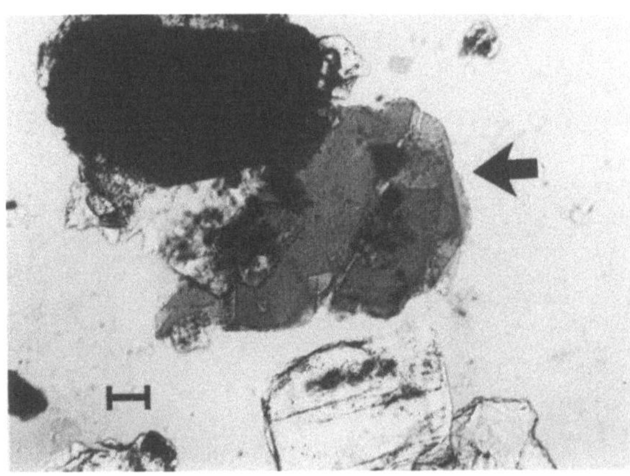

Fig. 76. Brown biotite mica flake (arrowed). Pliocene, south of Java Trench, Northeast Indian Ocean (9°46.53'S, 102°41.95'E). DSDP Leg 22, 211-8-1, 63 cm. Plane-polarised light, scale bar equals 0·05 mm.

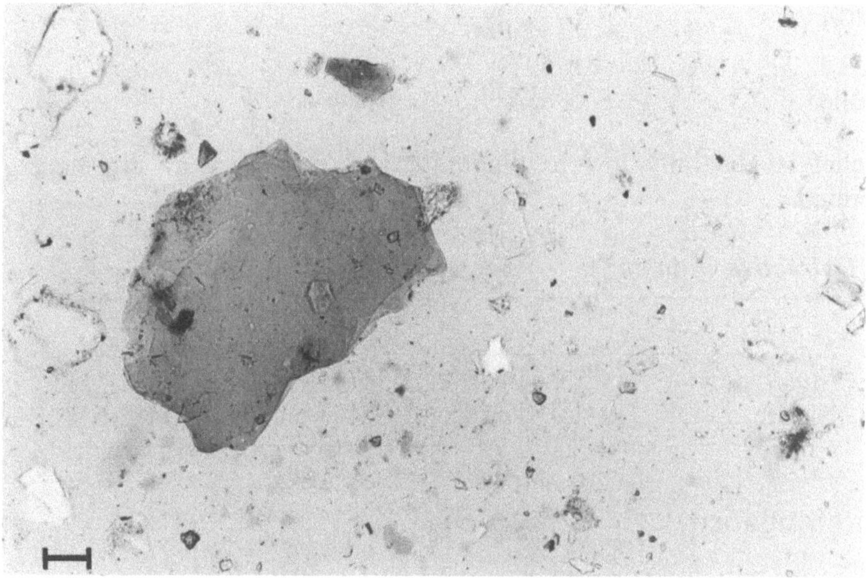

Fig. 77. Biotite mica, brown in transmitted light. Note the conspicuously large size of the flake compared with the associated material. Pleistocene, Bengal Fan, Indian Ocean (8°00.42′N, 86°16.97′E). DSDP Leg 22, 218-3-2, 130 cm. Plane-polarised light, scale bar equals 0·05 mm.

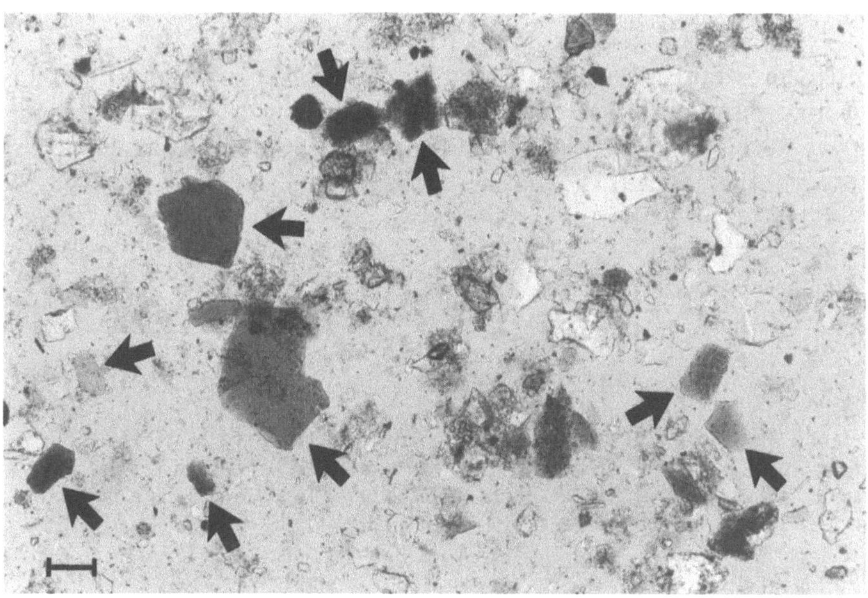

Fig. 78. Field of green and brown biotite flakes (arrowed). Pleistocene, Bengal Fan, Indian Ocean (8°00.42′N, 86°16.97′E). DSDP Leg 22, 218-3-2, 130 cm. Plane-polarised light, scale bar 0·05 mm.

Relief

Relief is generally low to moderate and may vary as the stage is rotated.

Refractive indices

	n_α	n_β	n_γ
Muscovite	1·556–1·570	1·587–1·607	1·593–1·611
Biotite	1·541–1·579	1·574–1·638	1·574–1·638

Birefringence

The flaky habit usually results in grains settling with {001} approximately parallel to the plane of the slide. Grains therefore usually present basal sections to the viewer. Basal sections of muscovite have low birefringence. Basal sections of biotite are dark coloured in plane-polarised light and practically isotropic in cross-polars.

Alteration

Muscovite is fairly resistant to weathering. Biotite has a lower stability and may show partial alteration to chloritic matter. Such grains may appear 'bleached'. Altered biotite flakes may sometimes resemble fish debris.

Crystal System

All micas are monoclinic.

Cleavage

Micas have perfect cleavage parallel to {001} yielding thin elastic plates.

Pleochroism

Although some micas are pleochroic in thin section, they are rarely seen to be so in smear slides. This is a result of the preferred orientation that most micas have in such slides—tending to lie with

{001} in the plane of the slide. Grains therefore show little, if any, variation in colour as the stage is rotated. In thin section, some varieties of muscovite are pleochroic from colourless to pale green. Biotite is pleochroic in shades of brown, red-brown or green and some varieties of chlorite are pleochroic in greens.

Extinction Angle

Extinction is generally parallel to the cleavage traces with basal sections being practically isotropic. Bent flakes may show wavy extinction.

Twinning

Twinning on the mica law (twin plane {110}, composition plane {001}) is fairly common, although rarely obvious in smear slides. However, such twinning may result in slight differences in interference colours or extinction.

Optical Character

Both muscovite and biotite have biaxial negative interference figures, although the isogyres of biotite's interference figure are so close, that the interference figure appears almost uniaxial. {001} flakes give good interference figures.

Occurrence/Source

Although micas usually comprise less than 1% of a sediment, their usually large grain size compared to associated material makes them conspicuous in smear slides. Micas are largely of continental derivation and so are usually present in shelf, slope and rise sediments. However, their flaky habit results in a hydraulic equivalence of much smaller particles and even sand-sized mica has transport characteristics similar to clay (Neiheisel, 1965). Consequently, micas can travel far from land and may be transported great distances by atmospheric or ocean currents. Sediments from areas adjacent to continental metamorphic or granitic terrains may show correspondingly high mica contents. Fluvial transport, coastal erosion, mass gravity flows and particularly aeolian transport all play major roles in transporting mica into the ocean realm.

Absence of mica in a terrigenous sediment indicates either that no mica was contributed to the sediment or that energy conditions were such that winnowing and/or by-passing of fines were important processes during sediment deposition. The presence of mica may indicate that conditions were such that there was little winnowing or by-passing of fines. There is a close correlation between mica content and gross energy levels. Doyle *et al.* (1968) assessed the relative importance of winnowing and by-pass processes on the southern United States Atlantic continental margin on the presence or absence of detrital mica.

Main Distinguishing Features

Flaky habit, transparency, colour, cleavage traces, good interference figures on {001} flakes.

Appearance as Loose Grains

Colourless, white, brown or green flakes and scales with pearly lustre. Such flakes often appear silvery in reflected light and may show iridescence due to incipient parting of cleavage planes within the flake.

Appearance in Thin Section

MUSCOVITE
Plane-polarised light: colourless or faintly coloured tabular crystals or scaly aggregates or shreds. Moderate relief.
Cross-polarised light: strong birefringence giving maximum interference colours of second order. Extinguishes parallel to the cleavage.

BIOTITE
Plane-polarised light: brown or green six-sided tabular crystals or lamellar aggregates. Moderate relief. Pleochroic in browns or greens. Pleochroic haloes around inclusions of radioactive minerals are common.
Cross-polarised light: strong birefringence but polarisation effects often masked by body colour. Extinction parallel to cleavage traces.

CHLORITIC MATTER
Plane-polarised light: greenish scaly masses and tabular plates. Moderate relief. Usually pleochroic in shades of green.
Cross-polarised light: birefringence generally weak but interference colours often anomalous (e.g. inky blue, dark brown, purplish-brown). Extinction is parallel to the cleavage traces in most chlorites.

MICRONODULES (Fe–Mn OXIDES AND HYDROXIDES)

Authigenic

Micronodules form part of a continuum of iron and manganese oxides and hydroxides ranging from microscopic flecks to macroscopic nodules, concretions and crusts. They generally consist of aggregates of various poorly crystalline hydroxides and oxides, among which the iron oxyhydroxide, goethite, is common.

Form/Shape

Often spherical but sometimes irregular in shape (Figs 79–81). Some micronodules may have shapes suggestive of a biogenic origin, reminiscent of internal casts of foraminifera or faecal pellets. Foraminifera tests may form the cores of some micronodules (Bonte *et al.*, 1980).

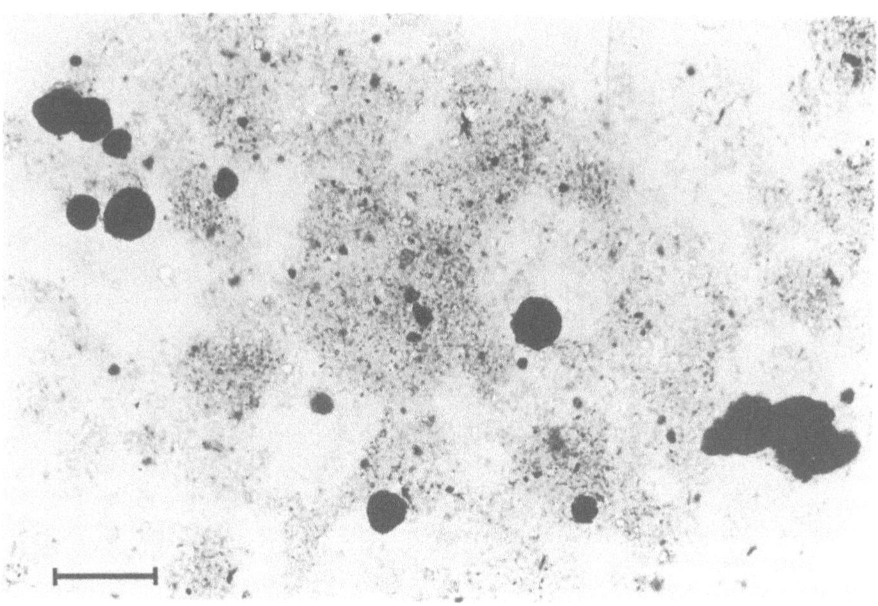

Fig. 79. Field of spherical to irregular micronodules with clay grade material. Pleistocene, Nares Abyssal Plain, Northwest Atlantic. Plane-polarised light, scale bar equals 0·05 mm.

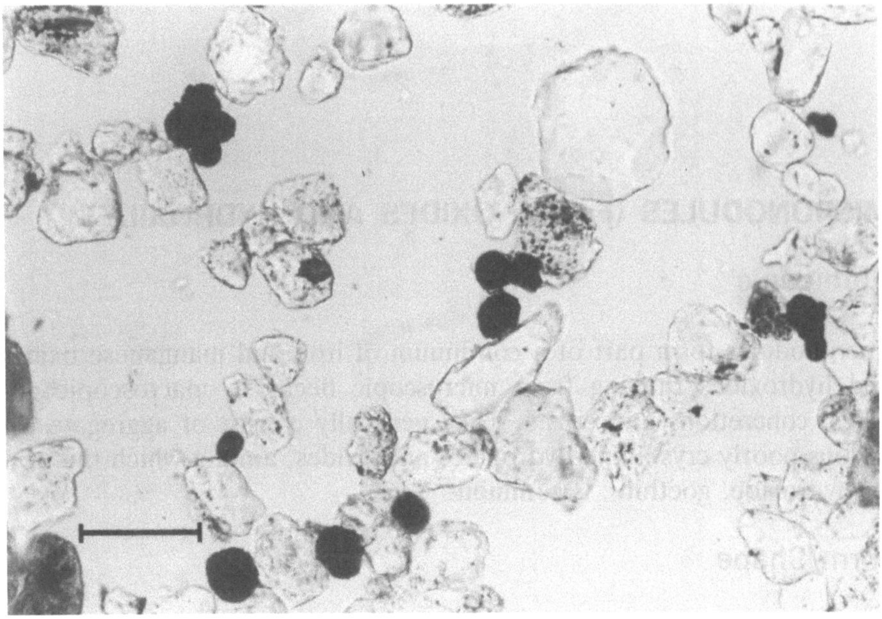

Fig. 80. Scatter of spherical to irregular micronodules. Most of the other material in the field of view is angular to subrounded quartz. Pleistocene, Madeira Abyssal Plain, Northeast Atlantic. Plane-polarised light, scale bar equals 0·05 mm.

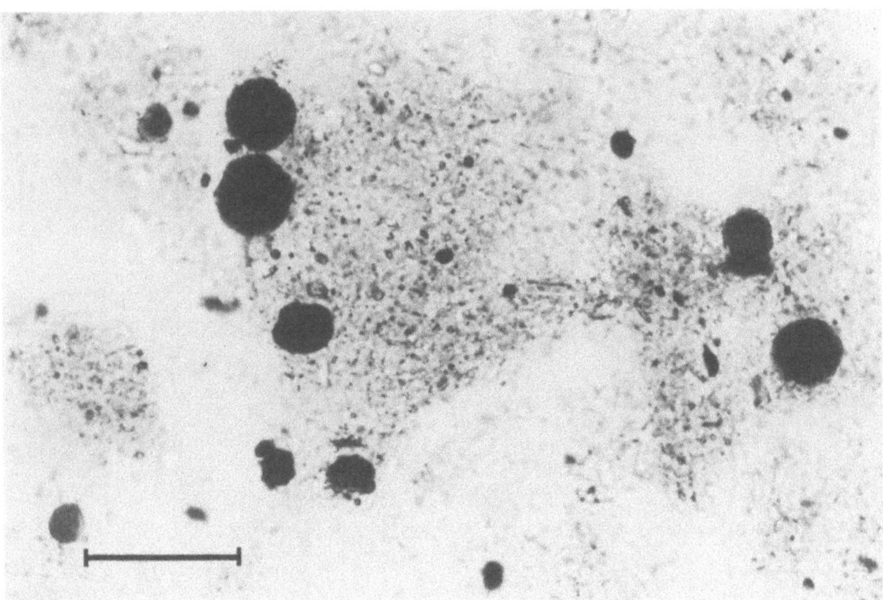

Fig. 81. Field of mainly spherical micronodules with clay grade material. Pleistocene, Nares Abyssal Plain, Northwest Atlantic. Plane-polarised light, scale bar equals 0·05 mm.

Grain Size Range

Very variable, but usually in the silt–sand size range (4–1000 μm).

Transparency/Colour in PPL

Always opaque.

Crystal System

The constituent iron and manganese oxides and hydroxides that comprise micronodules are mainly poorly crystalline and in some cases, amorphous.

Associated Minerals

Frequently associated with fish debris, altered volcanic rock fragments and sometimes rhodochrosite. Micronodules may occur in intimate association with rhodochrosite, as the two forms are the end members of a continuous transformation series with rhodochrosite ($MnCO_3$) transforming to manganese oxide grains with increasing redox potential (Borella & Adelseck, 1979) (Fig. 82).

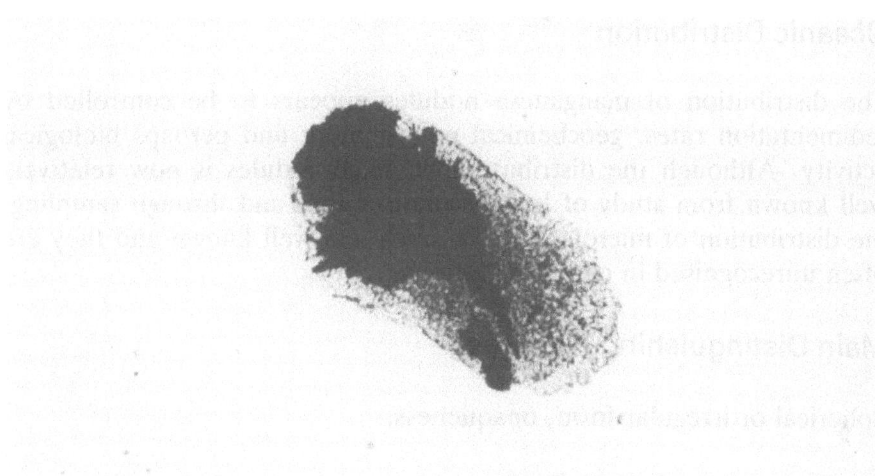

Fig. 82. Pink sand-sized rhodochrosite crystal showing transformation to a manganese micronodule. About 40% of the grain has been transformed to manganese oxide (opaque). ?Late Eocene, southern Bermuda Rise, Northwest Atlantic (25°06.63′N, 68°02.48′W). DSDP Leg 51, 417A-11-4, 130–132 cm. (Courtesy of Drs P. E. Borella and C. Adelseck and the Deep Sea Drilling Project.)

Occurrence

Micronodules may occur in both shallow-water and deep-sea sediments and may be distributed throughout the sediment column, although generally, they are more common near the sediment–water interface. They have been described from both lacustrine and marine deposits. The occurrence of both macro and micronodules is a function of several factors, notably sedimentation rates, the redox potential of the environment, the presence or absence of nucleating agents, the age and nature of the substrate, biological influences and the proximity of sources of elements (Cronan, 1980). Both local and regional variations in distribution will result from the interplay between these factors. Ferromanganese oxides will not precipitate in reducing environments and therefore will not be found in stagnant basins or on areas of the seafloor where reducing conditions prevail at the sediment surface.

Macronodule formation is most favoured by low sedimentation rates which prevent the growth of embryonic nodules being arrested through sediment burial. Some authors (e.g. Ku, 1977; Lalou *et al.*, 1980; Bonte *et al.*, 1980) favour a hydrothermal origin for at least some manganese nodules.

Oceanic Distribution

The distribution of manganese nodules appears to be controlled by sedimentation rates, geochemical environment and perhaps biological activity. Although the distribution of macronodules is now relatively well known from study of bottom photographs and through sampling, the distribution of micronodules is much less well known and they are often unrecognised in core descriptions.

Main Distinguishing Features

Spherical or irregular form, opaqueness.

Appearance as Loose Grains

Black, bluish-grey, greenish-grey or red-brown spherical or irregular grains whose surface texture may range from botryoidal and subvitreous to dull and porous. Broken grains may or may not show a concentrically layered structure with central nuclei. Individual grains may be crushed to powder under pressure from a needle.

Remarks/Notes

Ferromanganese deposits in the ocean reflect the fact that the redox potential and pH of seawater are such that iron and manganese, derived mainly through the weathering of marine and terrestrial volcanics, tend to oxidise and precipitate. The traditional view is that the formation of nodules takes place at the sediment surface as a precipitate from seawater or due to upward diffusion of reduced manganese from the sediment column (Bonatti & Nayuda, 1965), although Borella & Adelseck (1979) have provided evidence for a sub-surface *in situ* formation for some micronodules, through the transformation of rhodochrosite.

The literature on ferromanganese macronodules is vast. See Heath (1982) for a recent review, while detailed reviews of the mineralogy, distribution, geochemistry and economic aspects are contained in Glasby (1977). In contrast, the literature concerning micronodules is small.

PALAGONITE

Altered mafic volcanic glass

Palagonite varies in chemical composition, but Kerr (1977) gives the general formula SiO_2, Al_2O_3, Fe_2O_3, FeO, MgO, CaO, H_2O.

Form/Shape

Occurs as irregular flakes, semi-regular fragments and shard pseudomorphs (Figs. 83–85). Vesicular fragments may show a pseudo-oolitic structure due to the infilling of small vesicles (Fig. 86). May occur as marginal rims of varying thickness around glass fragments.

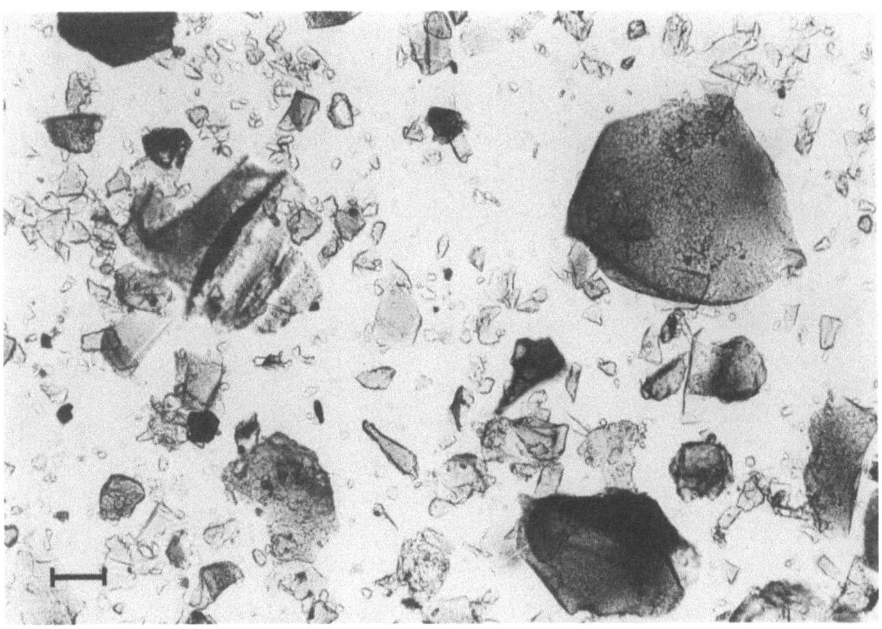

Fig. 83. Field of palagonitised volcanic glass, bright orange in plane-polarised light. Abyssal hill, Madeira Abyssal Plain, Northeast Atlantic. Plane-polarised light, scale bar equals 0·05 mm.

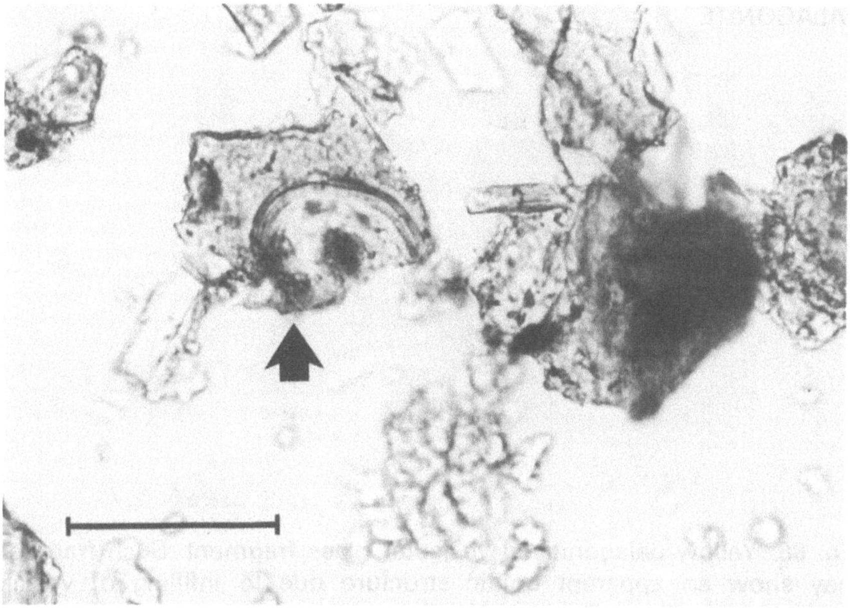

Fig. 84. Yellow palagonitised glass shards. Shard (arrowed) shows a vesicular embayment. Hyaloclastite, Kings Trough, Northeast Atlantic. Plane-polarised light, scale bar equals 0·05 mm.

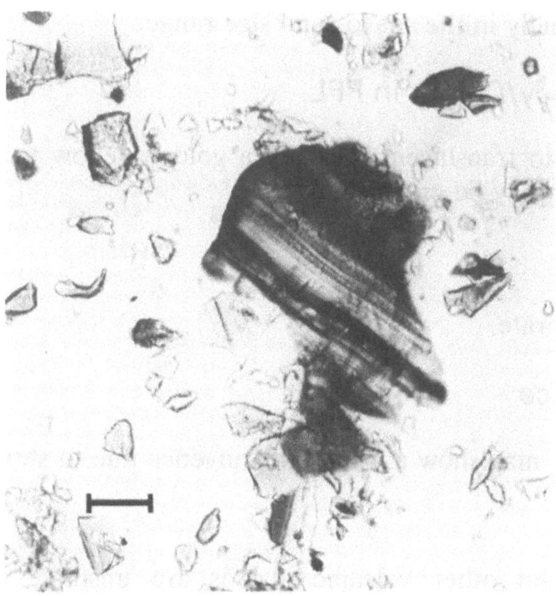

Fig. 85. Bright orange palagonitised glass fragment. The banded structure was probably originally caused by flowage. Abyssal hill, Madeira Abyssal Plain, Northeast Atlantic. Plane-polarised light, scale bar equals 0·05 mm.

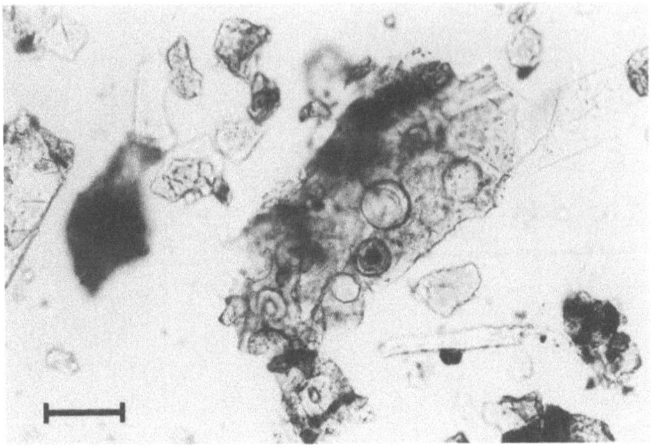

Fig. 86. Yellow palagonitised vesicular glass fragment. Such fragments may show an apparent oolitic structure due to infilling of vesicles. Hyaloclastite, Kings Trough, Northeast Atlantic. Plane-polarised light, scale bar equals 0·05 mm.

Grain Size Range

Variable, usually in the silt to sand size range.

Transparency/Colour in PPL

Transparent to translucent. Usually a golden yellow to reddish-brown in colour but may be greenish.

Relief

Low to moderate.

Birefringence

Often nil but may show a weak birefringence due to strain.

Alteration

Palagonite and other volcanic glasses are unstable in the marine environment and are readily transformed to smectites and zeolites.

Crystal System

Palagonite is generally non-crystalline (X-ray amorphous).

Cleavage

None.

Pleochroism

Absent.

Associated Minerals

Commonly associated with volcanic glass, pyroxene and plagioclase feldspar. Constituents lost from mafic glass during palagonitisation may be precipitated as zeolites and calcite.

Occurrence

Generally most common in marine volcanic provinces, where it may be formed through the hydration and granulation of the glassy rims of pillow lavas; or where there has been interaction between effusing lava and seawater, and through the hydration of hyaloclastites.

Source

Hydration of hyaloclastites, hydration and granulation of pillow lava rims, alteration of basaltic ash falls.

Indicator

Basaltic volcanism.

Main Distinguishing Features

Colour, form, nil or weak birefringence, association.

Appearance in Thin Section

Plane-polarised light: yellow to yellowish-brown glassy fragments or rims to glass shards. May also occur massive and may show a pseudo-oolitic structure due to the filling of small vesicles. Low to moderate relief. Cross-polarised light: nil or weak birefringence.

Notes/Remarks

Basaltic and felsic glasses generally yield different alteration products, particularly during the early stages of diagenesis. Palagonite appears to be derived mainly from mafic glasses while felsic glasses, being more resistant to devitrification, do not so readily alter to palagonite. (Sholze & Mulfinger, 1959; Marshall, 1961; Bonatti, 1965).

PYRITE

Authigenic/detrital
FeS_2
Iron disulphide

Form/Shape

Pyrite shows a variety of different morphological forms (Fig. 87). It can occur as:

(a) Euhedra—either as single cubes, aggregates of cubes or pyritohedra (Figs 87 and 88). Octahedra are less common.

(b) Framboids—these are spheroidal grains composed of pyrite microcrystallites. They are frequently seen within microfossil tests (Figs 89 and 90). According to Schallreuter (1984), framboidal pyrite was probably the most common form of pyrite seen in sediments recovered by the Deep Sea Drilling Project.

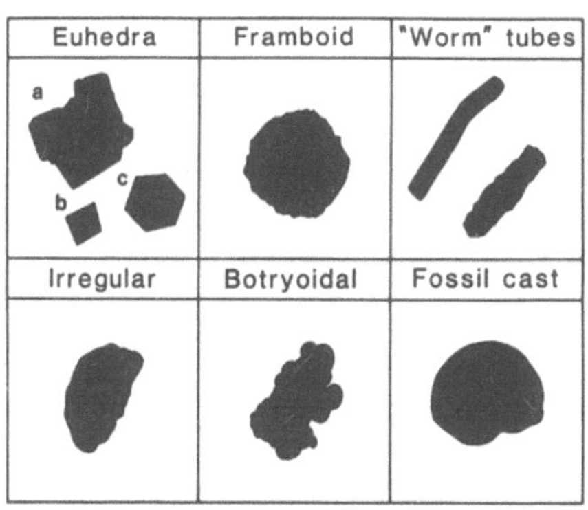

Fig. 87. Diagram showing the different morphological forms that can be taken by pyrite. These are euhedra, e.g. aggregates of cubes (a), octahedra (b) and pyritohedra (c); framboids (spheroidal grains composed of pyrite microcrystallites, frequently seen within microfossil tests); 'worm' tubes (straight or curved tubes which may have granular surfaces); irregular masses; botryoidal aggregates and fossil casts, in this case, of a small planispiral mollusc (after Siesser, 1978). The diagram shows examples of each morphological form as they would appear in plane-polarised light.

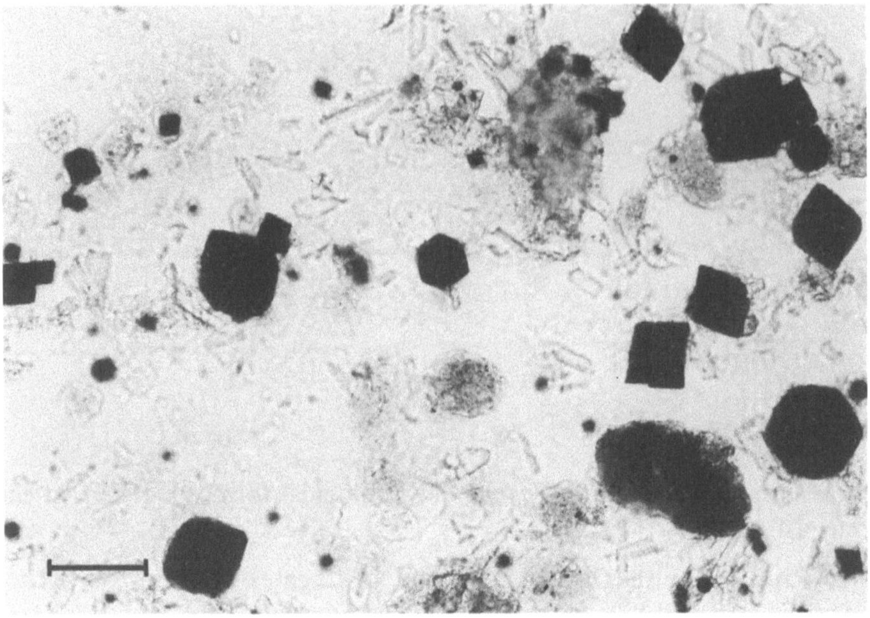

Fig. 88. Euhedral pyrite, forms include discrete octahedra and pyritohedra. Clumps of clay and colourless prismatic zeolite crystals are also present in the field of view. Late Miocene, western edge of the Sigsbee Deep, Gulf of Mexico (23°47.80'N, 94°46.09'W). DSDP Leg 10, 90-5-4, 100 cm. Plane-polarised light, scale bar equals 0·05 mm.

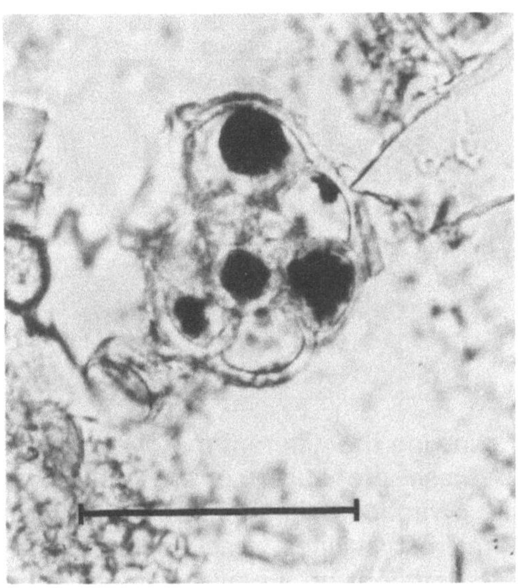

Fig. 89. Framboidal pyrite seen within the chambers of a foraminifer test. Middle Eocene, western flank of Bermuda Rise, Northwest Atlantic (30°50.39'N, 67°38.86'W). DSDP Leg 1, 6-4-2, 50 cm. Plane-polarised light, scale bar equals 0·05 mm.

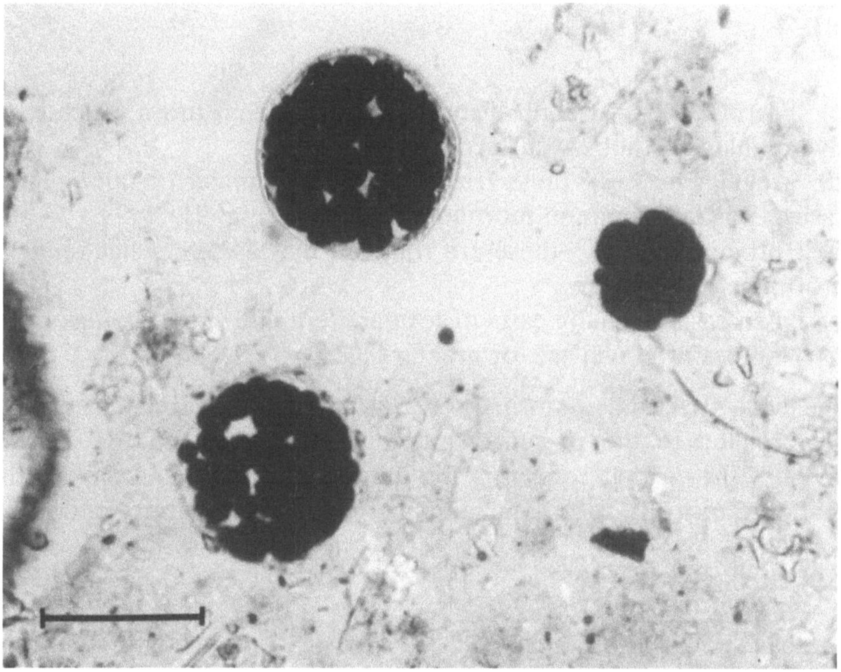

Fig. 90. Clusters of framboidal pyrite enclosed within diatom frustules. Late Pleistocene/Holocene, Angola Basin, Southeast Atlantic (19°11.26′S, 9°23.17′E). DSDP Leg 75, 530B-2-3, 13 cm. Plane-polarised light, scale bar equals 0·05 mm.

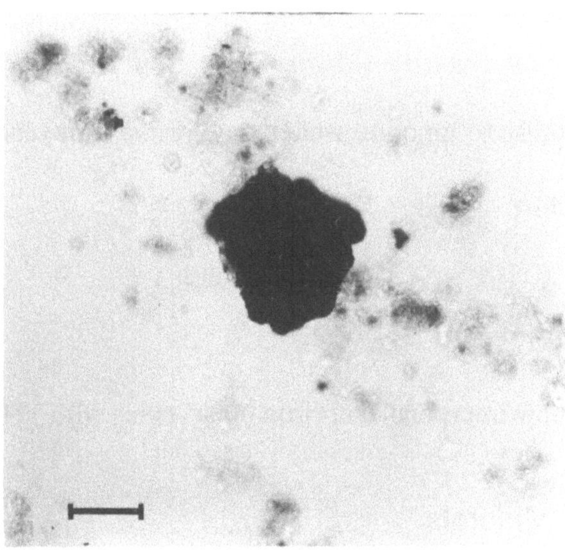

Fig. 91. Irregular mass of pyrite. Lower Cretaceous, lower continental rise, west of Cape Agulhas, South Africa (35°03.97′S), 15°26.91′E). DSDP Leg 40, 361–28-6, 33 cm. Plane-polarised light, scale bar equals 0·05 mm.

(c) 'Worm' tubes—straight or more usually curved tubes which have finely granular or knobbly surfaces.

(d) Irregular masses—these frequently have granular or microplanar surfaces. This is a common morphological form (Fig. 91).

(e) Botryoidal grains—these are rounded masses somewhat reminiscent of bunches of grapes.

(f) Fossil casts—usually casts of foraminifera and small molluscs.

(g) Nodules of gravel size or greater.

The wide variation in form shown by pyrite in sediments may be largely related to the presence and form of organic matter. Berner (1969) has shown that concentration gradients in sediments favour the migration of dissolved iron and sulphur towards an organic mass.

Grain Size Range

Pyrite can show a wide range of grain size. It can range from very fine silt-sized disseminated grains (0·003 mm/4 μm) to centimetre-sized concretions and nodules.

Transparency/Colour in PPL

Pyrite is always opaque.

Alteration

Pyrite can oxidise to limonite which may form a thin veneer on grains.

Crystal System

Cubic.

Twinning

Pyrite may show interpenetrant 'iron cross' twins with {110} as the twin axis.

Associated Minerals

Pyrite may be associated with glauconite and dolomite. Donnelly and Nalli (1973), Criddle (1974), Siesser and Rogers (1976) and Siesser (1978) report associations of pyrite and gypsum. This association may be considered unusual in that one mineral is a sulphide and the other a sulphate. Siesser (1978) infers that the association may occur in localised reducing environments which have a pH at least temporarily low enough to dissolve some calcareous biogenic debris. This would release calcium which could combine with already available sulphate ions to form gypsum, once its solubility product was exceeded.

Occurrence/Origin

Authigenic pyrite is associated with anaerobic or partly anaerobic environments. It forms during shallow burial through the reaction between detrital iron minerals and hydrogen sulphide produced by the bacterial reduction of interstitial dissolved sulphate (Berner, 1970; 1984). Initially, the iron compounds produced are black, finely disseminated iron monosulphides which are converted to the disulphide (pyrite) through reaction with elemental sulphur, probably produced through oxidation of hydrogen sulphide. According to Berner, most sedimentary pyrite formation occurs beneath partly aerobic water. Oxygen may be introduced into the sediment through bioturbation and perhaps storm and current agitation, allowing the formation of elemental sulphur (Siesser, 1978).

Authigenic pyrite has been recovered from both extremely anaerobic sediments and from apparently normally oxygenated open ocean sediments. However, even in such normally oxygenated sediments there must be localised microenvironments of oxygen depletion to allow anaerobic bacteria to begin the sulphate-reduction process. Examples are probably provided by the interiors of microfossil tests and places where organic matter has been buried such as burrows. High pyrite abundances are frequently associated with relatively high sedimentation rates (which prevent complete oxidation of the more easily decomposable organic matter near the sediment surface).

Main Distinguishing Features

Opacity, form, light brassy appearance under reflected light.

Appearance of Loose Grains

Single cubes, aggregates, framboids, fossil casts, irregular masses and nodules of variable size. Usually brassy yellow with a metallic lustre but may appear rusty brown or tarnished due to surface oxidation. Euhedral or subhedral grains may show striated crystal faces.

Appearance in Thin Section

Always opaque, even in the thinnest sections. Single crystals of euhedral pyrite will show square, rectangular, octahedral, triangular or even hexagonal outlines. Brassy yellow in reflected light with a metallic lustre.

Notes/Remarks

Whereas pyrite may be relatively common, its polymorph marcasite is rare in marine sediments. Both form in reducing conditions. Pyrite can form in weakly acidic, neutral and alkaline environments. In contrast, marcasite requires acidic conditions for formation (Kraus *et al.*, 1959) and these are unusual in marine sediments. However, marcasite may occur in sapropelic shales and other sediments having low pH values. Seisser (1978) reports occurrences of marcasite in DSDP cores from the Angolan continental slope (Sites 364 and 365) at horizons with interstitial pH values of around 6·5. The marcasite occurred as botryoidal aggregates and clusters of tabular or pyramidal crystals, either orientated irregularly to each other or radiating from a common centre. Like pyrite, marcasite is opaque and brassy yellow to greyish yellow in reflected light with a metallic lustre, although the surface is frequently oxidised to limonite.

QUARTZ

Detrital
Silicon Dioxide
SiO_2

Form/Shape

Well-rounded to very angular grains, often of moderate sphericity (Figs. 92–95). Fluid and mineral inclusions are common (Fig. 95). Grains may be single crystals or polycrystalline, comprising of a number of crystals per grain (best seen in cross-polars) and may show secondary outgrowths. Quartz bipyramids, characteristic of volcanic quartz are rare, although they may occur in some volcaniclastic sands. Grain form can be an important indicator of provenance.

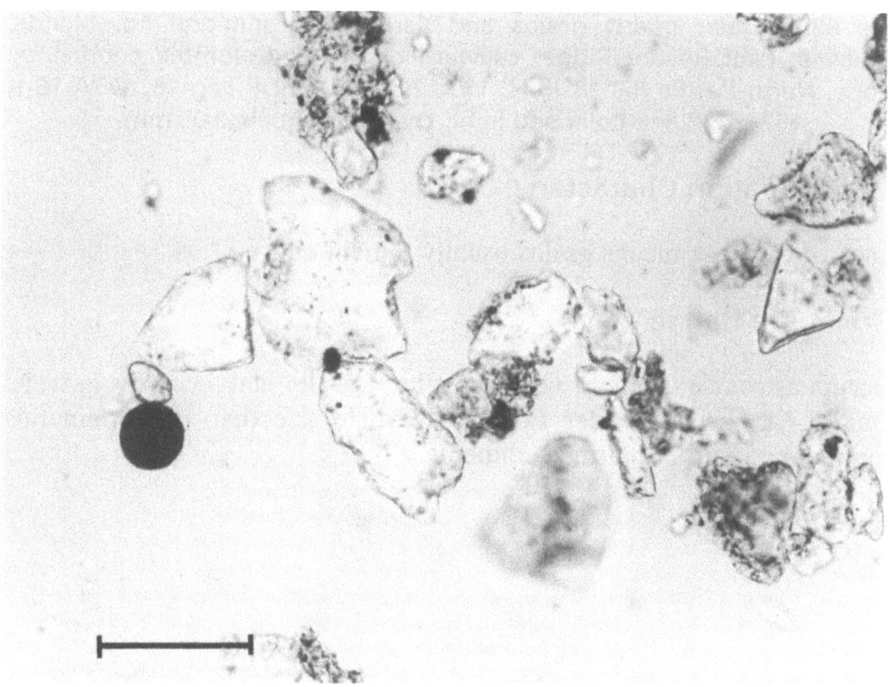

Fig. 92. Field of angular quartz grains. A spherical micronodule (left) is also in the field of view. Pleistocene, Madeira Abyssal Plain, Northeast Atlantic. Plane-polarised light, scale bar equals 0·05 mm.

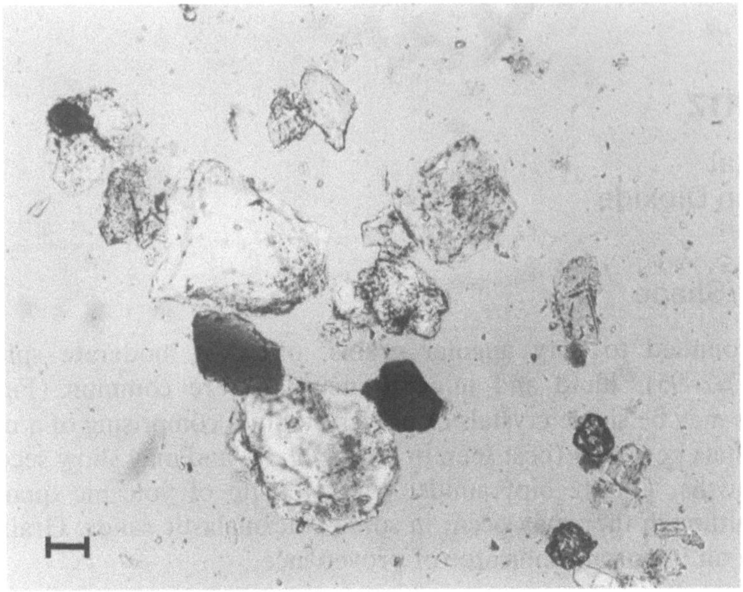

Fig. 93. Angular quartz grains and dark green anhedral hornblende. Pliocene, Paul Revere Ridge, seaward of British Columbia continental slope, North Pacific (50°28.18′N, 130°12.30′W). DSDP Leg 18, 177A-16-1, 57 cm. Plane-polarised light, scale bar equals 0·05 mm.

Fragmentation Characteristics

Lack of cleavage means grains usually remain entire.

Grain Size Range

Occurs as coarse sands to very fine silts. Locally may occur as gravels. Quartz has a much wider range of particle size than other minerals commonly found in marine sediments.

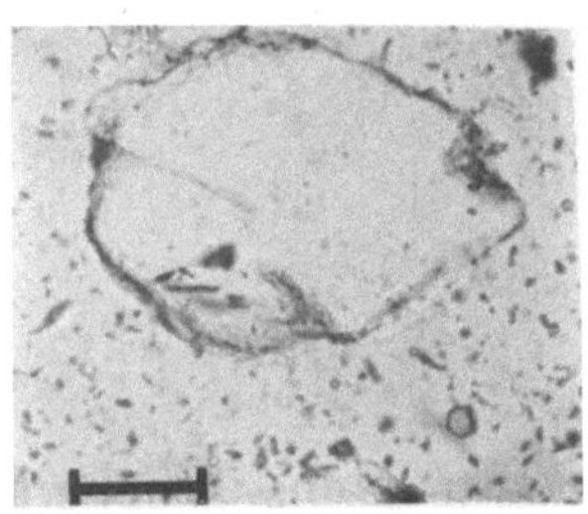

Fig. 94. Sub-rounded quartz grain. Pleistocene, Tyrrhenian Sea, Mediterranean. Plane-polarised light, scale bar equals 0·05 mm.

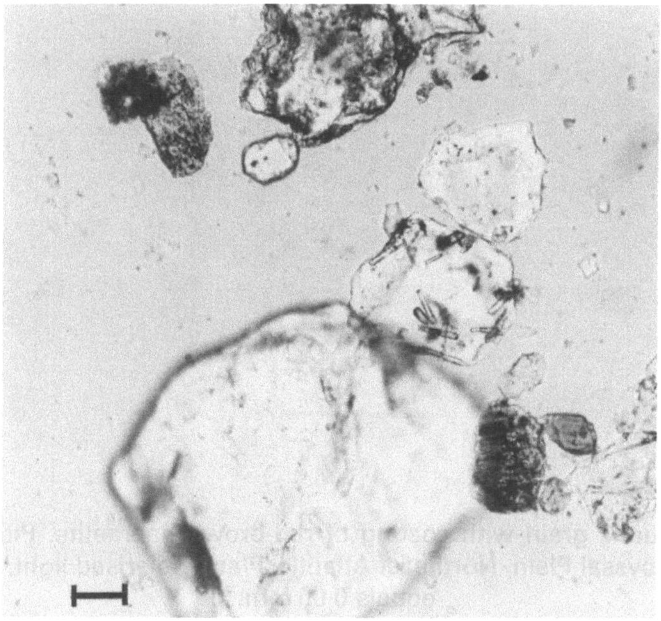

Fig. 95. Angular quartz, one grain containing crystallites. Pliocene, Paul Revere Ridge, seaward of British Columbia continental slope, North Pacific (50°28.18′N, 130°12.30′W). DSDP Leg 18, 177A-16-1, 57 cm. Plane-polarised light, scale bar equals 0·05 mm.

Transparency/Colour in PPL

Transparent although grains may appear slightly turbid. Grains are colourless, though aeolian grains derived from arid areas may have a coating of red-brown haematite (Fig. 96). Inclusions are usually sharply defined.

Relief

Low ($n_\omega = 1{\cdot}544$, $n_\varepsilon = 1{\cdot}553$).

Birefringence

Weak (0·009). Therefore grains show low interference colours, typically first order white, grey or yellow.

Alteration

None.

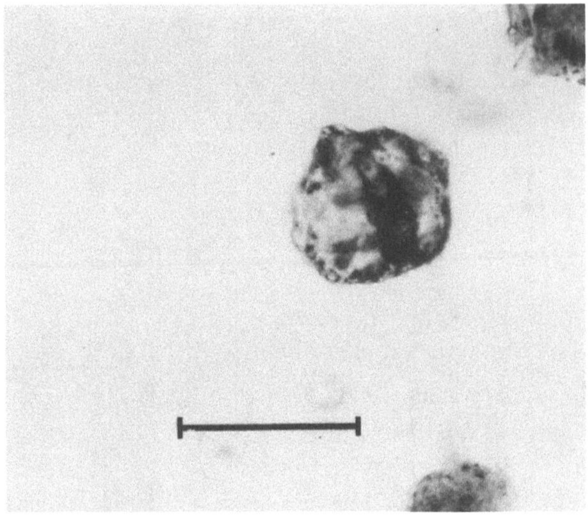

Fig. 96. Quartz grain with coating of red-brown haematite. Pleistocene, Madeira Abyssal Plain, Northeast Atlantic. Plane-polarised light, scale bar equals 0·05 mm.

Crystal System

Trigonal.

Cleavage

Absent.

Pleochroism

Absent.

Extinction

Parallel in euhedral crystals. Basal sections are dark in all positions. Irregular and wavy extinction, caused by strain, may be seen.

Twinning

Although twinning is common in quartz, it rarely shows in smear slides or thin section.

Optical Character

Basal sections at 30 μm give uniaxial positive interference figures without rings. However, interference figures of thick sections (greater than 1 mm) often have weak or hollow centres due to rotary polarisation.

Group Variability

Both Krynine (1946) and Folk (1980) have devised classifications of quartz types. Krynine's genetic classification of quartz types relates quartz character (specified by presence or absence of vacuoles, microlites and consideration of morphology and extinction) to ultimate origin. Folk's classification is purely empirical, based on extinction and inclusions, and does not make any genetic assumptions. Statistical analysis is then used to infer grain origin. The schemes may have some value in provenance studies, particularly for turbidite sands.

Occurrence

Quartz is ubiquitous in detrital sediments and can be found throughout the marine realm, although there is generally a marked decrease in grain size away from the continental slope and rise.

Oceanic Distribution

Distribution maps of quartz in surface sediments of the Atlantic Ocean are presented by Kolla *et al.* (1979). As quartz is nearly always of continental derivation, relatively low concentrations occur in the mid-ocean areas with pockets of high concentration towards the periphery. However, sediments off the mouths of major rivers (e.g. the Amazon, Plata, Congo and Niger) contain lower percentages of quartz than adjacent sediments (Kolla *et al.*, 1979). This reflects the fact that large rivers tend to carry proportionately large quantities of fine-grained clays rather than coarse sediment fractions. The ocean basins of the South Atlantic generally appear to contain a higher percentage of quartz than the equatorial basins. Kolla *et al.* (1979) attribute this to processes of supply, transport and redistribution by bottom or turbidity currents. The basins of the Atlantic contain relatively high concentrations of quartz compared to the other oceans. This probably reflects the relatively high ratio of the areas of the drainage basins on the adjacent continents to the area of the ocean, as compared to the corresponding ratios for other oceans (Windom, 1976). Distribution maps (see Kolla *et al.*, 1979, p. 265) show relatively high concentrations of quartz in North Atlantic sediments at high latitudes and this probably reflects the extensive granitic terrains exposed on adjacent continents. Fluvial transport and ice-rafting are probably the main transport mechanisms in this region.

Quartz distribution in the surface sediments of the Pacific Ocean have been studied by Rex & Goldberg (1958), Windom (1969) and Molina-Cruz & Price (1977). These studies showed a strong latitudinal variation of quartz in deep-sea sediments which correlates with the distribution of arid regions on the surrounding land. In general, North Pacific deep-sea sediments show quartz concentrations similar to those of the Atlantic, while those of the South Pacific are poorer in quartz. This probably reflects the ratios of the land drainage areas to those of the ocean basins. Rex & Goldberg (1958) record the highest concentrations of quartz in the Central North Pacific, suggesting large aeolian input. This view is supported by Windom (1969) on analysis of the size distribution of quartz in deep-sea sediments with that in atmospheric dust originating from the Australian deserts.

The distribution of quartz in Indian Ocean sediments has been studied by Goldberg & Griffin (1970), who record wind transport as the predominant transport mechanism in the Arabian Sea, while in the Bay of Bengal and Arabian Basin, fluvial transport is more important.

Main Distinguishing Features

Lack of alteration, absence of cleavage and observable twinning. Low relief and weak birefringence. Rounded and/or angular form. Ubiquitous occurrence in detrital sediments.

Source

Almost exclusively of continental origin.

Transport Mechanisms

Quartz may reach the deep seafloor by any type of sediment gravity flow. Other transport mechanisms for quartz into deep-sea sediments include ice-rafting, aeolian deposition and fall-out from high altitude jet streams. Much fine-grained quartz is probably transported together with the aluminosilicate suspensoid in seawater from the continents to the areas of deposition (Arrhenius, 1963). In some ocean basins, winnowing by bottom currents may locally concentrate quartz.

Indicator/Provenance

Quartz bipyramids indicate a calc-alkaline volcanic source, while polycrystalline grains indicate derivation from metamorphic terrains. Quartz grains coated with red-brown haematite indicate aeolian transport from arid continental regions. Such quartz may be an important component of deep-sea sediments, especially in equatorial regions. Diester-Haass (1976) & Kolla *et al.* (1979) have shown that during the last glacial maximum, the Sahara desert expanded southward and that this is reflected in a widening of the area of Atlantic sediments containing high abundances of red-stained aeolian quartz. Basu (1985) discusses aspects of reading provenance from detrital quartz.

Appearance as Loose Grains

Glassy angular to rounded subequent grains in the sand–silt size range, having a vitreous lustre. Haematitic quartz grains (sometimes termed 'wusten quartz') are generally well rounded and have pitted surfaces. Silt-size grains are commonly elongated parallel to the *c* axis, probably the result of obscure prismatic and rhombohedral fracture.

Appearance in Thin Section

At 30 μm thickness, quartz shows a maximum of first order white interference colour with a slight tinge of yellow. Other optical properties as above.

Notes

Authigenic silica is an important constituent of deep-sea cherts and porcellanites and appears to be largely derived from the diagenetic alteration of biogenic silica (Bramlette, 1946; Heath & Moberly, 1971; and others).

Authigenic silica phases observed in deep-sea cherts and porcellanites include chalcedony, opal-*CT*, opal-*A*, opal-*C* and high temperature forms of cristobalite and tridymite. These show characteristic X-ray diffraction patterns (Jones & Segnit, 1971). Chalcedony and opal-*CT* are the two most common authigenic silica phases. Chalcedony is a cryptocrystalline form of quartz having submicroscopic pores. Under the microscope it shows a fibrous structure and has a lower refractive index than crystalline quartz. Opal-*CT* occurs commonly as spherical–elliptical bodies composed of intersecting platy euhedral–subhedral crystals, 5–30 μm in diameter, termed lepispheres. These are usually found in voids and cavities in cherts. It may also occur massive, as a cement or more rarely as casts of radiolarian tests (Kastner, 1981).

VOLCANIC GLASS

Volcanic/detrital

Volcanic glass in marine sediments takes the form of shards and fragments resulting from explosively disrupted pumice, rapid quenching of molten lava or through the erosion of glassy rocks such as pumice or glass scoria.

Form/Shape

Glass shards may show a wide range of morphology, forming a continuum between cuspate and lunate bubble wall shards and fragments of highly vesiculated glass (Figs 97–100). Bubble wall shards

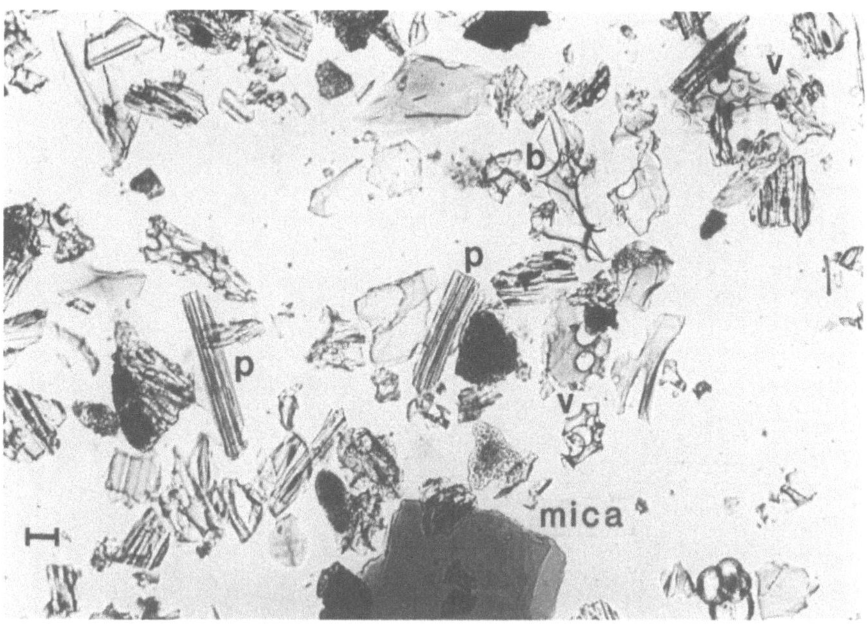

Fig. 97. Field of volcanic glass shards. A wide range of shard shapes are present, including semi-regular glass fragments packed with subparallel pipe vesicles (p), vesicular glass fragments (v) and multiwinged bubble wall shards (b). A large piece of biotite mica is also present in the field. Pleistocene, Tyrrhenian Sea, Mediterranean. Plane-polarised light, scale bar equals 0·05 mm.

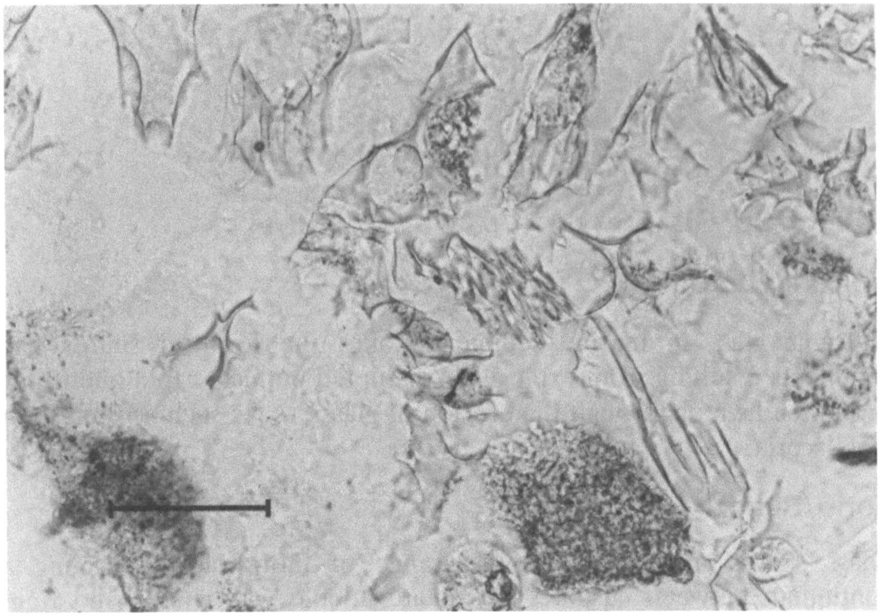

Fig. 98. Field of bubble wall shards, including cuspate and lunate forms. Pleistocene, Tyrrhenian Sea, Mediterranean. Plane-polarised light, scale bar equals 0·05 mm.

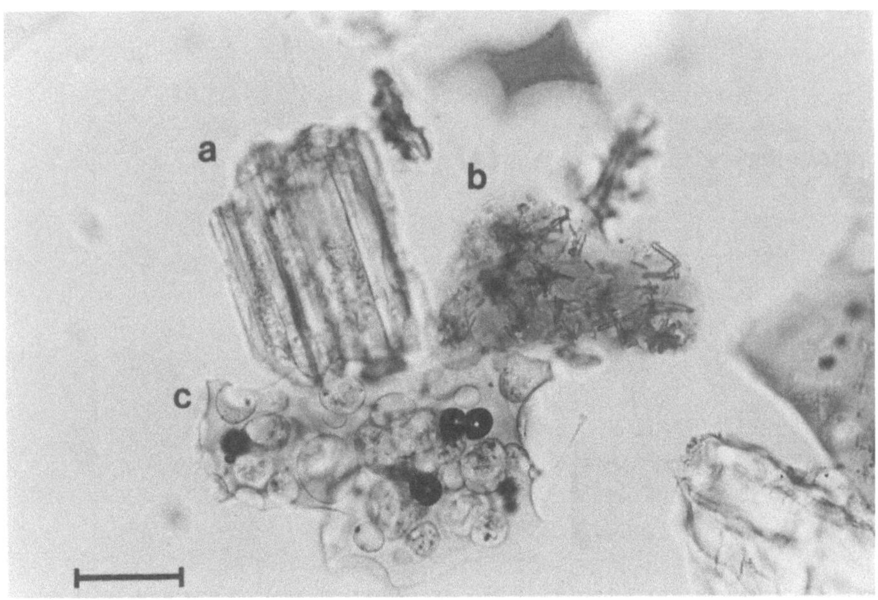

Fig. 99. (a) Semi-regular shard packed with pipe vesicles. (b) Coloured glass fragment containing abundant crystallites. (c) Vesicular glass fragment. Totally enclosed vesicles containing gas show up as high relief. Pleistocene, Tyrrhenian Sea, Mediterranean. Plane-polarised light, scale bar equals 0·05 mm.

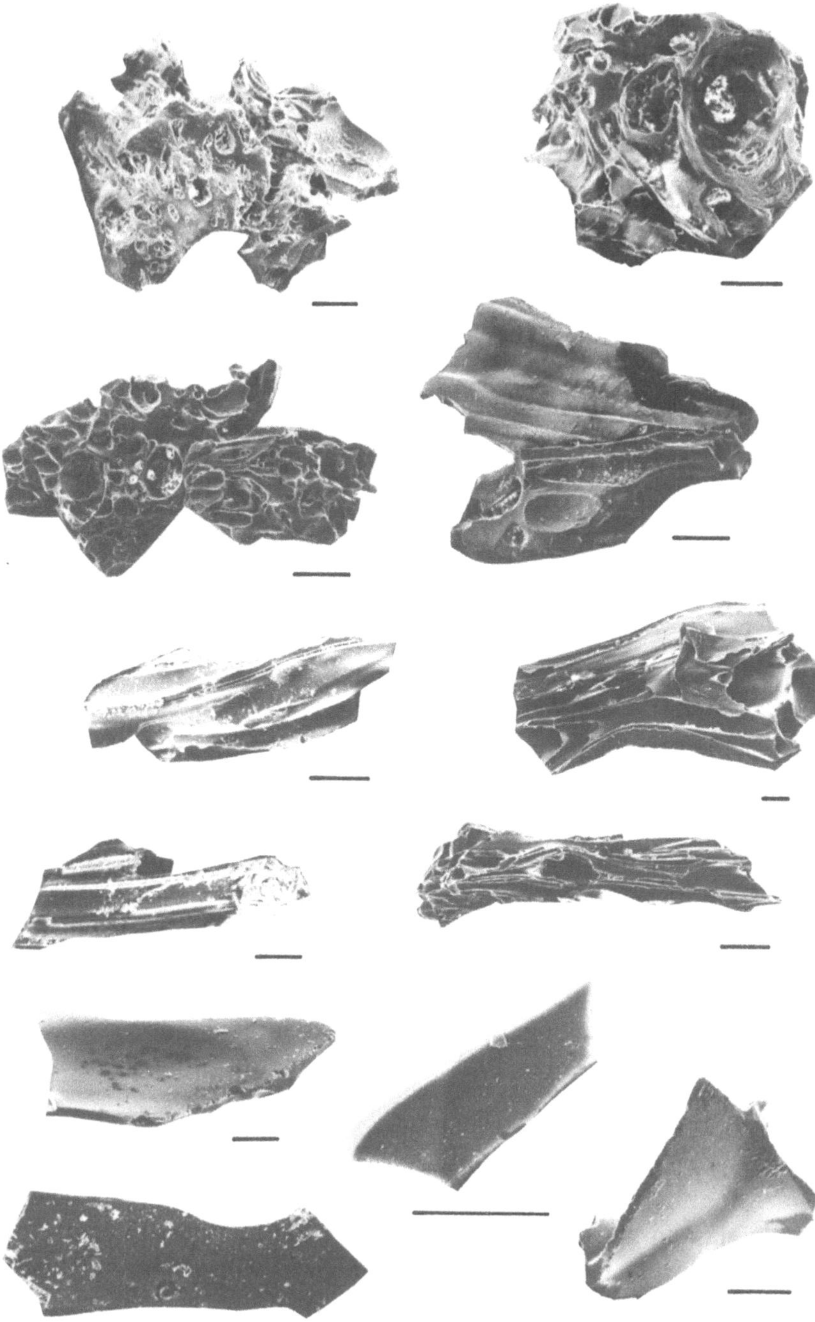

Fig. 100. Scanning electron micrographs of glass shards showing a range of shape from vesicular pumice (top), glass fragments containing abundant pipe vesicles (centre) to bubble wall shards (bottom). All Upper Pleistocene, south of Kings Trough, Northeast Atlantic. Scale bars are 30 μm and are to the bottom right of grains. (Courtesy of Mr Quentin Huggett.)

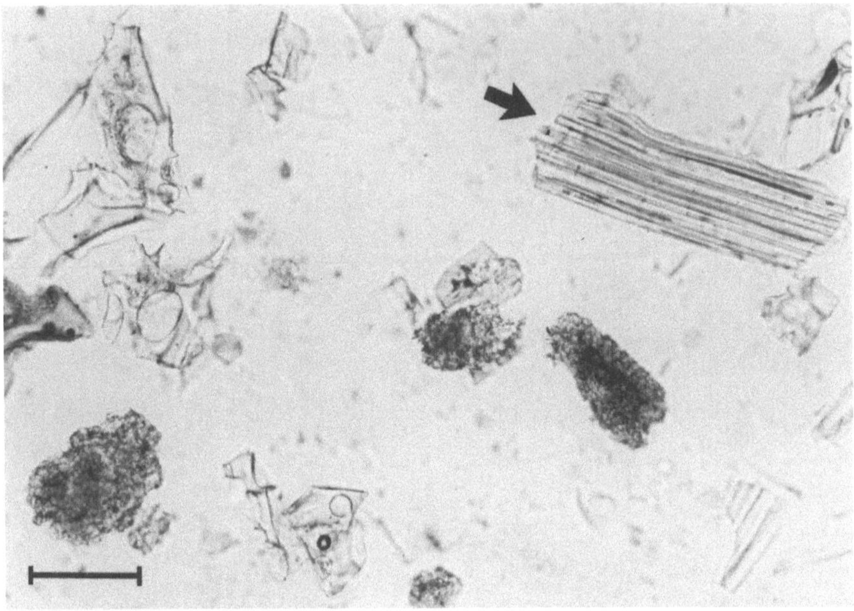

Fig. 101. Field of glass shards including a rectangular fragment containing numerous closely packed subparallel pipe vesicles (arrowed). Note the semi-regular form. Pleistocene, Tyrrhenian Sea, Mediterranean. Plane-polarised light, scale bar equals 0·05 mm.

comprise parts of the walls of broken vesicles and vary from curved plates to multi-winged fragments. They often show hemispherical or elliptical marginal embayments. Fragments may show a superficially fibrous structure reflecting the presence of large numbers of closely packed pipe vesicles (Fig. 101). Such fragments often have a semi-regular shape.

Grain Size Range

Volcanic glass in marine sediments varies from fine dust to coarse sand in size. However, most distinguishable glass will be in the fine silt (10 μm) to medium sand (500 μm) size range.

Transparency

Usually clear and transparent when fresh but may be obscured by vesicles, inclusions and/or incipient alteration. Darkly coloured glass may be translucent. Volcanic glass may contain mineral, gaseous or liquid inclusions (Fig. 102).

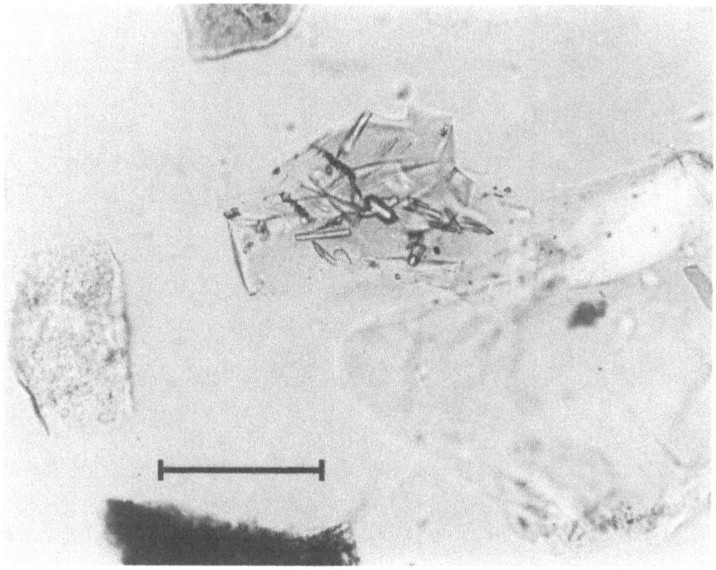

Fig. 102. Volcanic glass fragment packed with crystallites. Pleistocene, Tyrrhenian Sea, Mediterranean. Plane-polarised light, scale bar equals 0·05 mm.

Colour in PPL

May be colourless, green or brown. Coloured glasses are usually of a light shade. Generally, felsic glasses tend to be colourless while mafic glasses tend to be coloured. However, exceptions are numerous. Glass colour reflects composition and is probably controlled by transition metal content, especially that of iron oxides (Larsen & Berman, 1943).

Relief

Moderate, most volcanic glasses have refractive indices in the range 1·46–1·62. The refractive index of glass shards varies according to their composition (George, 1924) and there is a close correlation between refractive index and silica content (Williams *et al.*, 1954). Shards with a silica content of 78% or so, have a refractive index of 1·50. As the silica content decreases, the refractive index increases proportionately but the trend is slightly curved at silica contents of 55–65% (Fig. 103). Glass with a silica content of 50–60% has a refractive index of 1·57–1·60. However, refractive index not only varies with composition but also with subsequent alteration of the glass.

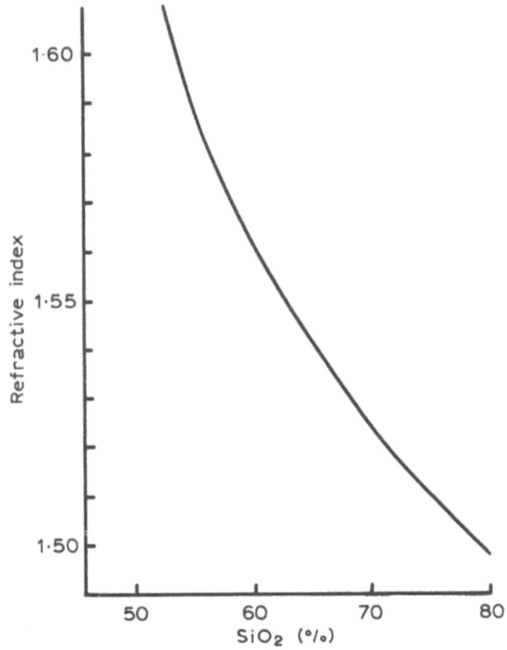

Fig. 103. Graph showing relationship between refractive index and SiO_2 content of volcanic glass shards (after Furuta & Arai, 1980).

The refractive index of volcanic glass shards primarily depends on the transition metal oxide content as the refractivity of transition metal oxides is higher than for other constituent oxides and silica (Larsen & Berman, 1943). The predominant transition metal oxide in volcanic glass is iron oxide, so the refractive index of glass shards closely correlates to total iron (Furuta & Arai, 1980). Refractive index increases with increasing iron oxide.

The refractive index of volcanic glass tends to remain constant for a single eruption. Consequently, refractive index may be used to determine approximate chemical composition and possible source of the glass (Mathews, 1951). Although it is possible for a number of eruptions to have glass of similar refractive index but be separated in space and time (Nayudu, 1964). Mellis (1954), Nayudu (1964), Ninkovich *et al.* (1964) and others have correlated ash layers in deep sea cores on the refractive index of the shards.

Birefringence

Fresh glass is isotropic, however, devitrification may result in low order interference colours.

Alteration

It is widely held that volcanic glasses are inherently unstable and will devitrify over time (e.g. Williams *et al.*, 1954). However, a number of instances of unaltered glass shards of considerable age are reported in the literature (e.g. Scheidegger, 1973) and even from sediments of Cretaceous age (Jones, 1973). Even so, some Pleistocene and Pliocene glasses rapidly devitrify, in some cases within a few hundred thousand years (Scheidegger *et al.*, 1978; Ninkovich, 1979). The devitrification of volcanic glass is discussed by Marshall (1961). Diffusion of water into the glass disrupts the glass structure and after such hydration relatively little energy is required to arrange the silica and alumina tetrahedra to form an ordered crystalline structure. Theoretically diffusion of water into the glass and hence devitrification is a slow process (Marshall, 1961), but minute fractures in the glass may quicken devitrification considerably. High temperatures also enhance the diffusion of water into the shards and thus devitrification can rapidly follow eruption. Bonatti (1965) concludes that glass produced through the direct interaction with seawater during a submarine volcanic eruption will rapidly hydrate and devitrify; while glass formed during subaerial eruptions and later transported into the ocean realm or produced during submarine eruptions in which the lava was somehow prevented from directly interacting with the seawater during cooling, will hydrate more slowly. This may partly account for the great variance in devitrification shown by volcanic glass.

Glass of different composition will hydrate at different rates, as diffusion coefficients are dependent on viscosity. Scholze & Mulfinger (1959) and Hawkins (1961) suggest that the activation energy for diffusion of water into natural glasses increases with increasing saturation (acidity). Hence felsic glass appears more resistant to devitrification than mafic glass. Hydration of volcanic glasses results in palagonite, zeolites and smectites. Alteration usually begins from the outside of the shard and works inward. Altered shards may show visible hydration rims.

Crystal System

Not applicable.

Cleavage

None.

Pleochroism

None.

Extinction Angle

Not applicable.

Associated Minerals

In older sediments, glass may occur with its alteration products, particularly palagonite, zeolites (especially phillipsite) and smectites. It may be associated with pyroxenes and other ferromagnesian minerals derived from volcanic rocks.

Terminology

Basaltic glass is known as tachylyte or sideromelane; jaspoid or hyalobasalt are less commonly used terms. Tachylyte containing olivine and pyroxene euhedra is sometimes called limburgite. Felsic volcanic glass is sometimes referred to as lechatelierite.

Occurrence

Volcanic glass in marine sediments may occur disseminated throughout the sediment or as megascopically distinguishable ash layers. Visible tephra layers are usually deposited within a few hundred kilometres from the parent subaerial eruption.

Oceanic Distribution

Both basaltic and felsic glass shards are widespread in marine sediments. Distribution patterns will be controlled by source regions, ocean currents and prevailing wind directions. Most of the tephra described from DSDP cores has been rhyolitic to andesitic in composition. Basaltic glass will be commonly associated with pillow lavas in areas of submarine eruptions such as mid-ocean ridges. Felsic shards from explosive terrestrial volcanism will enter the marine realm via atmospheric fallout from ashclouds. Felsic glass will most likely reflect crustal subduction or the eruption of highly differentiated magmas in other geological settings.

Source

Volcanic glass may originate during submarine volcanic eruptions through the granulation of pillow rims or during subaerial eruptions with later deposition from ash clouds. In general there is a decrease in average size of glass shards away from the source, but redistribution of the finer glass shards by water currents and input from other sources may cause a more erratic distribution pattern (Kidd, 1975). In several cases, the distribution of glass derived from volcanic ashfalls has been demonstrated to mirror the associated prevailing wind directions. Examples are eruptions of Hekla in Iceland (1947), Katmai in Alaska (1912) and Santa Maria in Guatemala (1902) as summarised by Lisitzin (1972).

Transport Mechanisms

Volcanic glass may be transported to the ocean realm through atmospheric fall-out, subaerial and coastal erosion, fluvial transport and mass gravity flows. In polar regions, ice-rafting is a further transport mechanism. The transport and deposition of tephra depends on many factors. These include:

—initial force of ejection;
—angle of ejection;
—height attained by the ashcloud;
—direction, velocity and turbulence of winds;
—the size, shape and density of the shards;
—water circulation and effects of currents.

Large particles will fall out of the ashcloud close to the source (Fisher, 1964; Slaughter & Hamil, 1970) but the finer shards may be widely dispersed by atmospheric currents. For shards under 40 μm in size atmospheric residence times increase greatly. Windom (1969) suggests that in certain areas of the ocean basins, up to 75% of the detrital material under 40 μm may be the result of atmospheric fall-out. The long atmospheric residence times for fine volcanic glass are illustrated by Mossop (1964) who reported the collection of atmospheric fine volcanic dust from the 1963 Agung eruption (Indonesia), 100 days or more after the eruption—enough time for the glass dust particles to encircle the Earth several times by transporting winds.

Vallier & Kidd (1977) conclude that volcanic ash can be more widely and effectively dispersed in the ocean than any other non-biogenic sedimentary component, because of the available methods of transport. Wind is the most effective agent for fine ash dispersal from explosive subaerial eruptions but fine ash may also be carried great distances by ocean currents (Fisher, 1965).

Shaw *et al.* (1974) have constructed a simple model for estimating palaeowinds and palaeo-explosivities from atmospherically transported volcanic glass in deep-sea sediments. From their model they have been able to predict fall-out patterns and associated particle size distributions for single eruptions and discuss the efficient design of piston-coring programmes to sample and define the shape of downwind eruption clouds and to estimate palaeo-explosivity.

Diagenesis

Basaltic and felsic glasses generally yield different alteration products, particularly during the early stages of diagenesis. Basaltic glass hydrates to yellow palagonite and the constituents lost from the glass during the palagonitisation process are precipitated as zeolites and calcite. Palagonite itself readily alters to smectites. Felsic glasses, on the other hand, are generally more resistant to devitrification and do not so readily alter to palagonite. Such glass usually first alters to zeolites and smectites (Hay, 1966). Relatively pure smectite deposits formed through the decomposition of volcanic glass are termed bentonites.

Main Distinguishing Features

Shape, common presence of vesicles and inclusions, absence of cleavage, lack of colour or pale coloration, isotropism in cross-polarised light. Comminuted biogenic silica may resemble shards but has a lower refractive index (Kerr, 1977).

Appearance as Loose Grains

En masse, colourless or faintly coloured volcanic glass has the appearance of silver sand. Even under relatively high magnification such glass appears yellowish and has a subvitreous lustre. Larger grains will appear colourless or faintly coloured, show clear shard morphology and have a vitreous lustre. In such grains, vesicles and bubble trains may be clearly observed. Fine grains will tend to stick together due to the interlocking of shard projections. Dark coloured glass may have a more resinous lustre and have greater opacity, but again should be distinguishable by its form.

Appearance in Thin Section

May appear colourless, grey or reddish in plane-polarised light. May show shard morphology or be massive. May be vesicular, perlitic and/or contain spherulites, microlites, crystallites and micropheno-crysts. In cross-polarised light, volcanic glass may show weak birefringence due to strain or alteration but when fresh, is isotropic. Frequently devitrified to indefinate alteration products.

Separation

Sand-sized glass fragments (over 62 μm) may be separated by hand-picking under a binocular microscope, as they are usually readily distinguishable by their form. Huang *et al.* (1975) have described a method for separating fine disseminated volcanic glass (under 62 μm) from sediments using heavy liquids. The reported efficiency of the technique ranges from 45 to 85% with increasing particle size. In their technique, glass shards may even be separated from siliceous microfossils by using liquids of specific densities.

Notes/Remarks

Besides refractive index, other properties of glass shards that may be useful in their correlation are presence or absence of vesicles, mineral inclusions and their shape. Electron microprobe analysis and atomic absorption are important techniques for studying chemical variability in volcanic glasses and are hence useful in provenance studies (see, for example, Rothe & Koch, 1978; Scheidegger *et al.*, 1978).

Glass shards have great potential for the dating of marine sediments through fission track dating (MacDougall, 1971). Windblown glass produced by subaerial explosive volcanism often has a relatively high potassium and uranium content, allowing the use of radiometric techniques as further dating tools. MacDougall (1971) describes fission track dating of glass shards in marine sediments that have also been dated by other methods. He concludes good agreement. After chemical etching, fission tracks will be visible using high power light microscopy and can be distinguished from bubbles, scratches, etc. by their pointed cone-like appearance.

ZEOLITES

Authigenic

Zeolites are a group of hydrous sodium potassium calcium aluminium silicates of diverse geological occurrence. They are commonly found in amygdales and other cavities in basic igneous rocks. Most zeolites found in sediments have been formed through the alteration of volcanic ash or other pyroclastic material. Although some thirty-three natural zeolites are known, only three—phillipsite, clinoptilolite and analcime are of any importance in marine sediments. Of these, phillipsite and clinoptilolite are by far the most common, analcime is next in abundance and the other zeolite species occur much more rarely. However, only phillipsite and sometimes clinoptilolite occur as sufficiently large crystals to be identified visually in smear slides, the other species usually being too fine-grained. Therefore only clinoptilolite and phillipsite are considered in detail here.

Zeolites tend to be particularly abundant in marine volcaniclastic deposits and they have been recorded as forming up to 80% of altered tephra horizons.

Clinoptilolite

Authigenic
Hydrous sodium potassium calcium aluminosilicate
$(Na, K)_4CaAl_6Si_{30}O_{72}.24H_2O$ (Sheppard quoted in Kastner & Stonecipher, 1978).

Form/Shape

Commonly occurs as very small subhedral to ragged grains or small euhedral platy crystals (Figs 104–106). More rarely, clinoptilolite can occur as large well-formed crystals, usually flattened on {101}. In co-occurences with phillipsite, it is nearly always the finer-grained material. However, when large crystals of phillipsite and clinoptilolite co-occur, it may be possible to distinguish between them, as clinoptilolite crystals tend to have blunter terminations compared to phillipsite (Fig. 107, *cf.* Fig. 109). Most clinoptilolite seen in smear slides will be very fine-grained and hence difficult to distinguish optically.

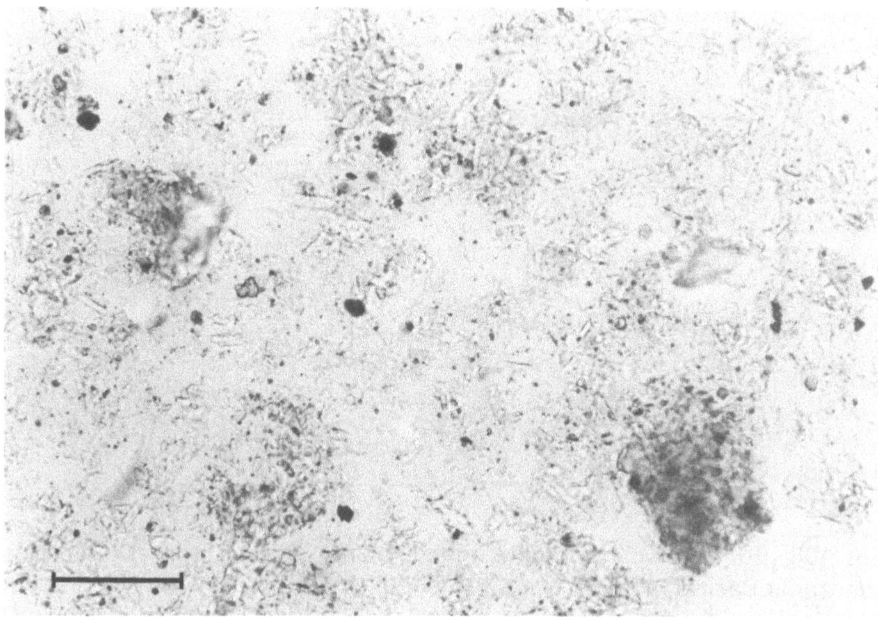

Fig. 104. Shreds of clinoptilolite with minor pyrite and clay grade material. Early Miocene, Venezuela Basin (14°30.69′N, 69°21.35′W). DSDP Leg 15, 150-5-Core-catcher. plane-polarised light, scale bar equals 0·05 mm.

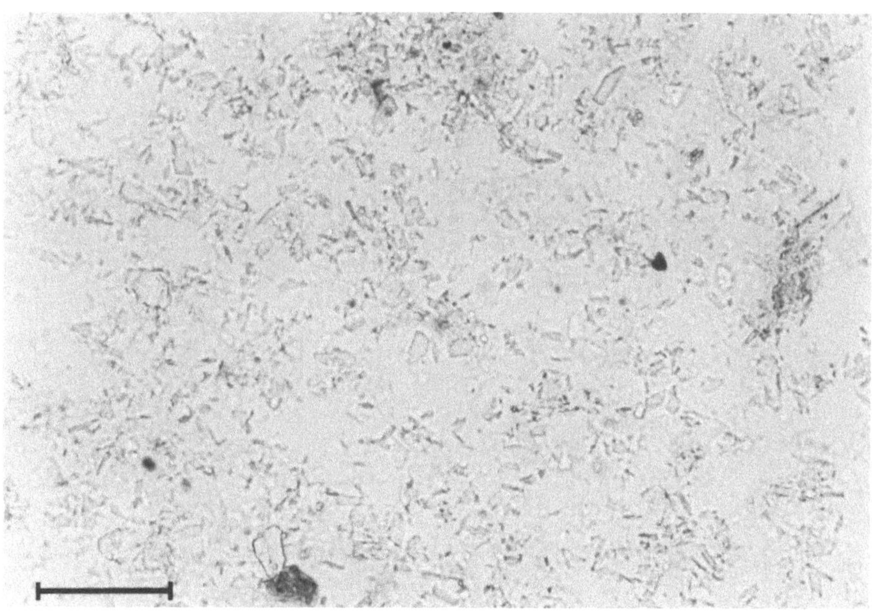

Fig. 105. Field of clinoptilolite. Most is in the form of small ragged shreds but a few small euhedral platy crystals are present. Early Miocene, Venezuela Basin (14°30.69′N, 69°21.35′W). DSDP Leg 15. 150-5-Core-catcher. Plane-polarised light, scale bar equals 0·05 mm.

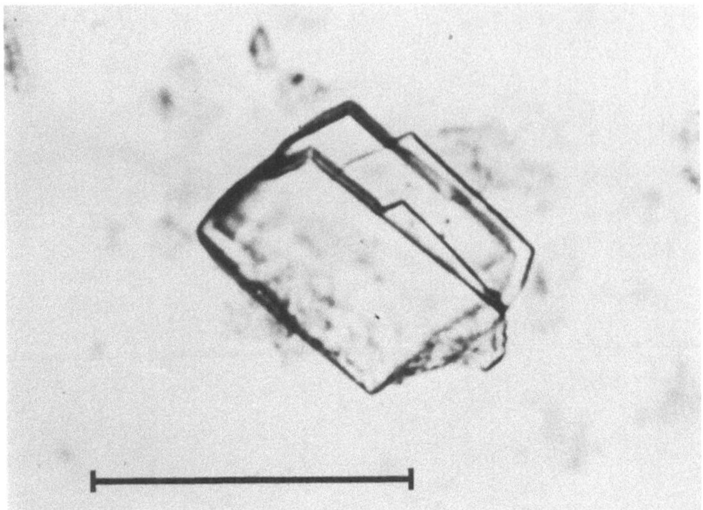

Fig. 106. Euhedral clinoptilolite crystal. Santonian, Upper Cretaceous, Venezuela Basin (14°30.69′N, 69°21.35′W). DSDP Leg 15, 150-9-1, 118 cm. Plane-polarised light, scale bar equals 0·05 mm. (Courtesy of Dr T. W. Donnelly and the Deep Sea Drilling Project.)

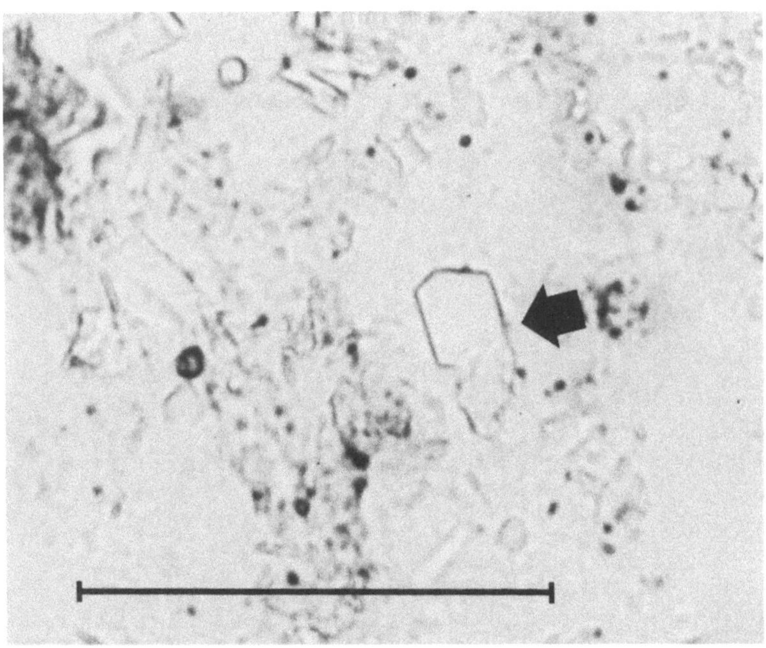

Fig. 107. Clinoptilolite crystal showing euhedral form (arrowed). Note that the crystal termination is less acute than in phillipsite. Early Miocene, Venezuelan Basin. DSDP Leg 15, 150-5-Core-catcher (14°30.69′N, 69°21.35′W). Plane-polarised light, scale bar equals 0·05 mm.

Grain Size Range

Most grains and crystals fall into the 2–45 μm size range.

Transparency/Colour in PPL

Usually transparent and colourless.

Relief

Moderate. Mean refractive index: 1·480–1·486 (Boles & Wise, 1978).

Birefringence

Anisotropic, weak birefringence.

Crystal System

Monoclinic.

Pleochroism

Absent.

Associated Minerals

Clinoptilolite is associated with calcareous and opal-*CT*-rich sediments (Kastner & Stonecipher, 1978). Minerals that may occur associated with clinoptilolite are calcite, chert, opal, siderite and clay minerals (primarily detrital smectites and illites in younger sediments and palygorskite and illites in older sediments) (Iijima, 1978; Kastner & Stonecipher, 1978; Stonecipher, 1978).

Occurrence

In contrast to phillipsite, clinoptilolite appears more common in older sediments (Stonecipher, 1978). It is least common in surface and shallowly-buried sediments but increases in abundance at depths below 700 m. Further, whereas phillipsite is most common in clays, clinoptilolite is more common in calcareous or calcareous/siliceous sediments (Kastner & Stonecipher, 1978). However, in some brown clays clinoptilolite may be remarkably abundant. Pimm *et al.* (1971) report a sample of Late Eocene/Early Oligocene brown clay from DSDP Site 59 in the western Pacific comprising 76% clinoptilolite.

Oceanic Distribution

In surface sediments, clinoptilolite appears more common in marginal areas of the ocean than phillipsite (Cronan, 1980). However, it has also been described from the central Pacific (Stonecipher, 1976; Petzing & Chester, 1979), from the Atlantic (particularly off North and South America) and from surface sediments off South Africa, Madagascar and Australia in the Indian Ocean (Kolla & Biscaye, 1973; Petzing & Chester, 1979; Cronan, 1980). Iijima (1978) in a study of deep-sea zeolite distribution based on X-ray diffraction data on samples from early DSDP sites concludes that zeolite distribution is distinct in each oceanic region. He found clinoptilolite to be the most common zeolite in the Atlantic but to be equally as abundant as phillipsite in the Pacific Ocean. Yet on the Pacific margins, clinoptilolite is far more abundant than phillipsite as in the Atlantic. However, Iijima found that phillipsite was especially abundant in pelagic sediments of Pacific marginal seas associated with island arcs. In the Indian Ocean, clinoptilolite is again the dominant zeolite. Petzing & Chester (1979) reviewed the distribution of both clinoptilolite and phillipsite in Upper Miocene to Recent sediments in the Pacific and Atlantic Oceans; based on data extracted from early DSDP volumes. They noted a close correlation with the present day pattern of global volcanicity. They see the correlation between zeolite types and both sediment lithologies and sedimentation rates as largely coincidental.

Origin

Most workers consider that clinoptilolite is derived through the low-temperature low-pressure alteration of felsic volcanic glass (Hay, 1966; Sheppard & Gude, 1968; 1969; and others) while the alteration of basaltic glass results in phillipsite. Sheppard & Gude (1969) suggest that clinoptilolite may form directly from felsic glass by solution precipitation in moderately alkaline and saline environments. A direct association of clinoptilolite with silicic volcanic glass is reported by Cook & Zemmels (1972) from DSDP Site 84. Petzing & Chester (1979) propose that it is the Si/Al ratio of the parent glass that controls whether phillipsite or clinoptilolite will form, clinoptilolite formation being favoured by a high Si/Al ratio. They explain the greater abundance of phillipsite in younger sediments by the fact that basic glasses alter more quickly than acidic ones which are more resistant to devitrification. Consequently, clinoptilolite is less abundant in younger sediments. However, it seems clear that clinoptilolite can also form without volcanic precursors. Berger & von Rad (1972) and Mélières (1979) report the diagenetic formation of clinoptilolite through the infilling and replacement of radiolarian tests. Thus, in some instances, abundance of clinoptilolite may indicate abundance of silica in surface waters and hence siliceous plankton productivity. Boles & Wise (1978) suggest that clinoptilolite may form through the alteration of phillipsite with increasing age.

Phillipsite

Hydrous potassium sodium aluminosilicate
$K_{2\cdot8}Na_{1\cdot6}Al_{4\cdot4}Si_{11\cdot6}O_{32}.10H_2O$ (Stonecipher, 1977; 1978; Kastner, 1981).

Form/Shape

Usually occurs as elongated euhedral to subhedral prismatic crystals. These may be discrete (Figs 108 and 109), show sector/cruciform

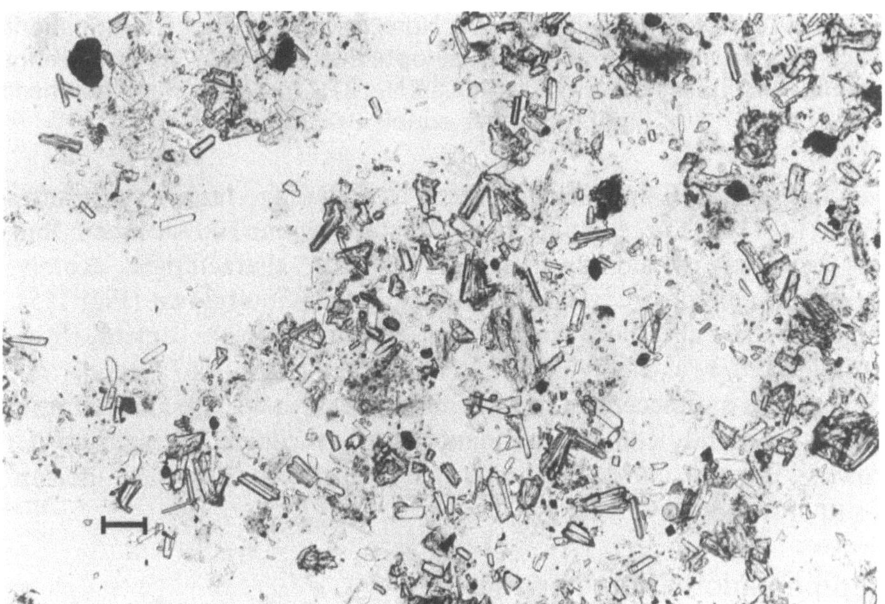

Fig. 108. Field of phillipsite crystals. ?Early Neogene, Kings Trough, Northeast Atlantic (44°02.80′N, 21°47.03′W). Plane-polarised light, scale bar equals 0·05 mm.

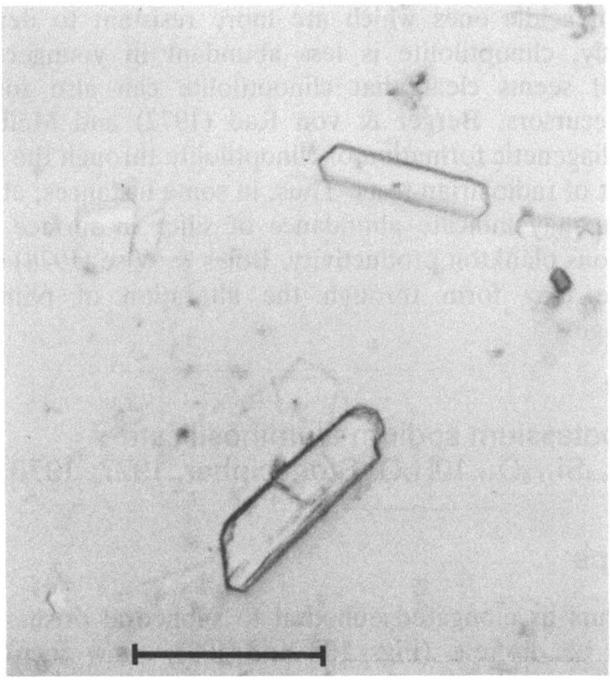

Fig. 109. Discrete phillipsite prisms. Note the characteristic terminations which are more acute than in clinoptilolite. ?Early Neogene, Kings Trough, Northeast Atlantic (44°02.80′N, 21°47.03′W). Plane-polarised light, scale bar equals 0·05 mm.

twinning (Figs 110 and 111) or occur as radiating clusters or rosettes (Figs 112–114). Large grains may be multi-twinned about their long axes to form pseudo-hexagonal prisms with characteristic acutely-angled terminations. Crystals may be complexly intergrown (Fig. 115). Phillipsite has also been reported as occurring as a cement (Morgenstein, 1967), as cavity and fracture fillings (Rex, 1967; Bass *et al.*, 1973) and as a replacement of plagioclase (Bass *et al.*, 1973). Phillipsite crystals frequently contain abundant inclusions indicating rapid crystal growth. Bonatti (1963) suggests that well-formed crystals indicate in-situ growth.

Fragmentation Characteristics

Broken crystals are common in smear slides. Complete sector twins are rarely seen, although singly-terminated crystals, representing twin components, are common.

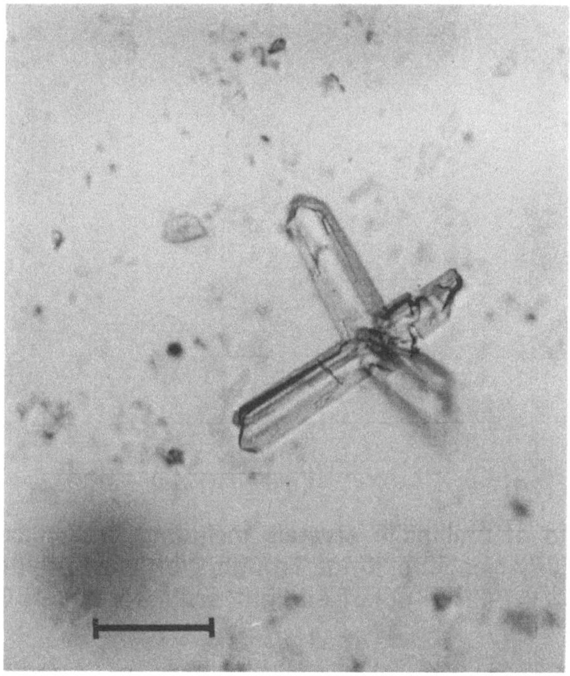

Fig. 110. Broken phillipsite cruciform twin. ?Early Neogene, Kings Trough, Northeast Atlantic (44°02.80′N, 21°47.03′W). Plane-polarised light, scale bar equals 0·05 mm.

Grain Size Range

Generally in the range 8 to 250 μm with most crystals falling into the size range 10 to 120 μm; although occasional occurrences of much larger crystals have been reported (Sheppard *et al.*, 1970).

Transparency/Colour in PPL

Transparent and colourless to yellowish. The yellowish colour is due to numerous inclusions of matrix components and to some extent to surface coating of Fe–Mn oxyhydroxide (Kastner, 1981).

Relief

Moderate to high. Mean refractive index: 1·481–1·486 (Kastner & Stonecipher, 1978).

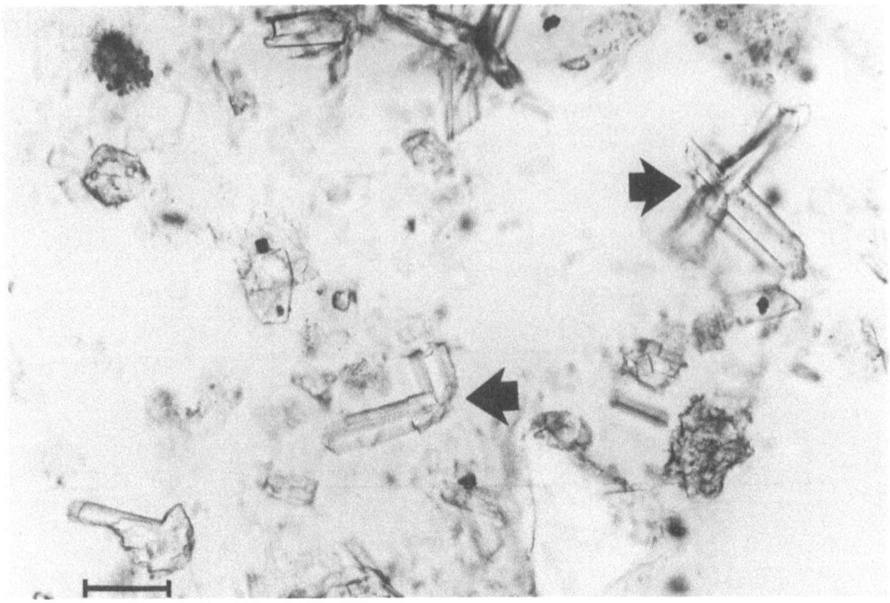

Fig. 111. Field of phillipsite crystals including broken cruciform twins (arrowed). ?Early Neogene, Kings Trough, Northeast Atlantic (44°02.80′N, 21°47.03′W). Plane-polarised light, scale bar equals 0·05 mm.

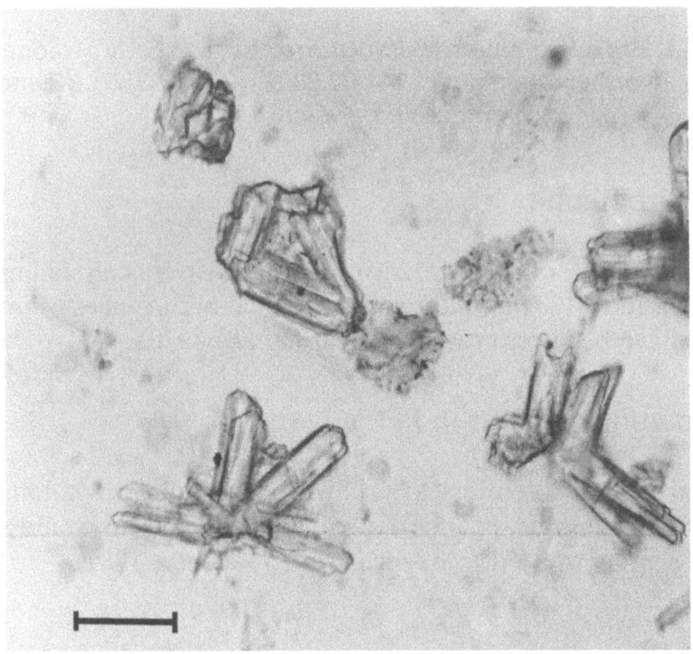

Fig. 112. Partial rosettes and complexly intergrown phillipsite prisms. ?Early Neogene, Kings Trough, Northeast Atlantic (44°02.80′N, 21°47.03′W). Plane-polarised light, scale bar equals 0·05 mm.

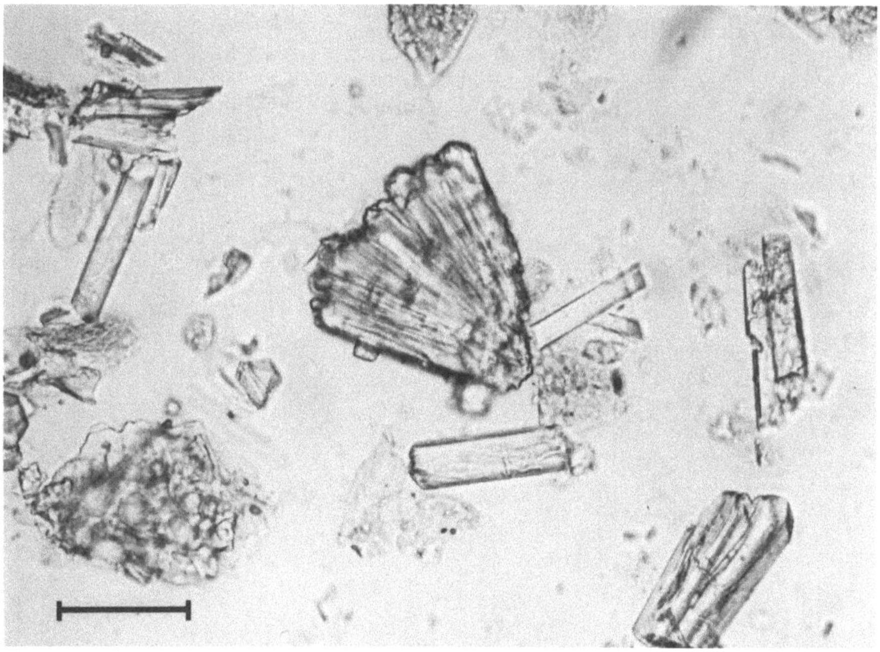

Fig. 113. Partial rosette (above centre) and bladed crystals of phillipsite. ?Early Neogene, Kings Trough, Northeast Atlantic (44°02.80′N, 21°47.03′W). Plane-polarised light, scale bar equals 0·05 mm.

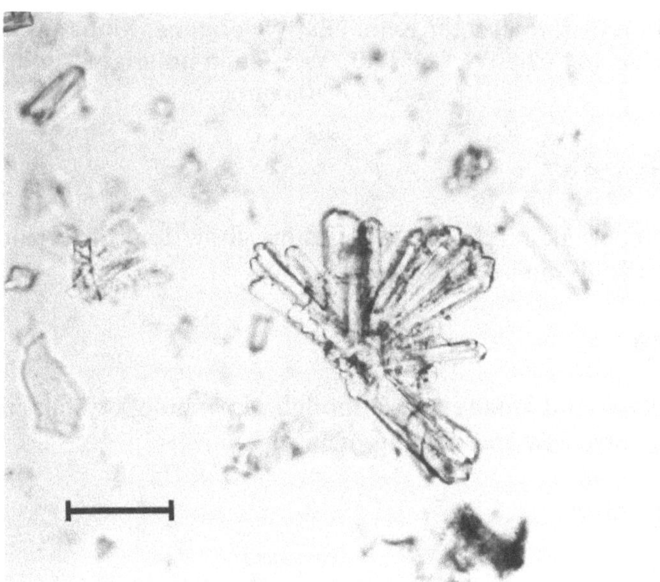

Fig. 114. Partial phillipsite rosette. ?Early Neogene, Kings Trough, Northeast Atlantic (44°02.80′N, 21°47.03′W). Plane-polarised light, scale bar equals 0·05 mm.

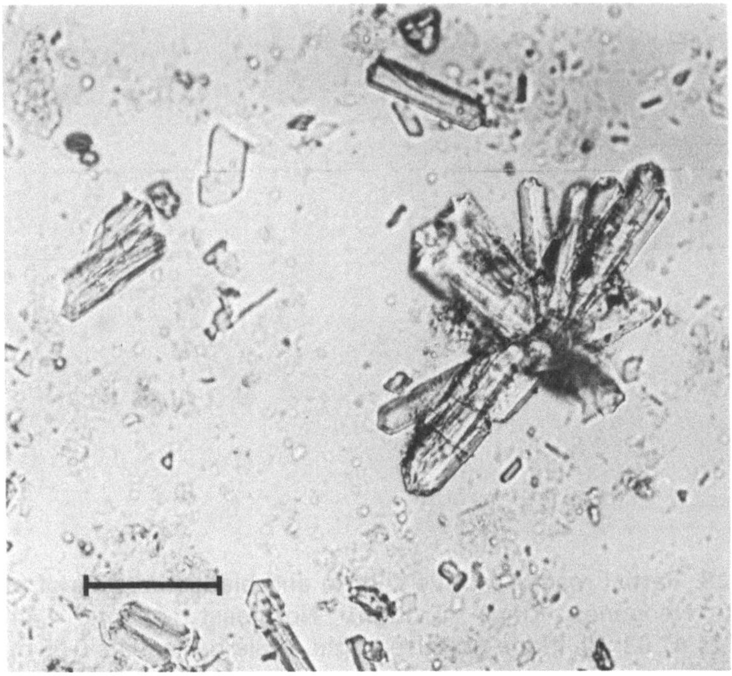

Fig. 115. Scatter of phillipsite crystals including a large complexly intergrown cruciform/sector twin. ?Early Neogene, Kings Trough, North-east Atlantic (44°02.80′N, 21°47.03′W). Plane-polarised light, scale bar equals 0·05 mm.

Birefringence

Anisotropic, weak birefringence. Grains show low order interference colours, typically first order greys.

Alteration

Usually clear and unaltered, although downhole crystals may show increasing corrosion and etching of faces.

Crystal System

Monoclinic.

Pleochroism

Absent.

Extinction Angle

Prismatic grains show practically straight extinction.

Twinning

Sector twinning, sometimes complex, is common. Simple cruciform twins may occur.

Group Variability/Related Species

Harmotome is a zeolite with the same structure as phillipsite but with barium as the principal cation (Rinaldi *et al.*, 1974). Harmotome appears to be relatively uncommon. Arrhenius & Bonatti (1965) report the occurrence of harmotome as microcrystals dispersed through aggregates of volcanic glass hydration products. They suggest that it may represent an early hydration product which is usually subsequently overgrown by phillipsite. Although uncommon, harmotome has been described mainly from areas of the Pacific with low sedimentation rates and is usually associated with phillipsite (Kastner & Stonecipher, 1978; and others). Harmotome is unlikely to be distinguished in smear slides.

Associated Minerals

Often associated with other volcanogenic material, most frequently palagonite and basaltic volcanic glass. Plagioclase, barite and ferromanganese oxides and hydroxides may also be associated with phillipsite. The clay-fraction of phillipsite-bearing sediment is usually rich in smectite.

Occurrence

Phillipsite is widespread over large areas of ocean floor, especially regions with low sedimentation rates and below the calcium compensation depth (Cronan, 1980). It is especially common in altered tephras. It has been reported from red clays, siliceous and calcareous sediments and terrigenous muds but Stonecipher (1976), in a statistical analysis, recorded the following sequence of decreasing phillipsite abundance as a function of lithology: clay > volcanic > calcareous > siliceous. The occurrence of phillipsite appears to be a function of geological age; it is far less common in pre-Miocene sediments (Kolla & Biscaye, 1973; Stonecipher, 1976).

Oceanic Distribution

Current data suggests that phillipsite is more abundant in Pacific and Indian Ocean sediments than in those of the Atlantic (Kolla & Biscaye, 1973; Stonecipher, 1976). Petzing & Chester (1979) record high concentrations from the Pacific and Bonatti (1963) has recorded sediments containing over 50% phillipsite on a carbonate-free basis from the central and southern Pacific. In the Indian Ocean, the main occurrences so far recorded are from the Central Indian and Wharton Basins and from parts of the Mid-Indian Ocean and Ninety-East Ridges (Kolla & Biscaye, 1973). In the Atlantic Ocean, reported instances are sparse. However, phillipsite-rich sediments have been described from the Cape Verde Basin, from south of the Sohm Abyssal Plain (Petzing & Chester, 1979) and from Kings Trough (Kidd *et al.*, 1982).

Origin Source

Present consensus is that phillipsite is derived from the extreme alteration of small basaltic glass shards on the seafloor, through a solution–precipitation step, probably smectite or palagonite (Honnorez, 1978; Iijima, 1978; Kastner & Stonecipher, 1978; Kastner, 1981; and others). However, Petzing & Chester (1979) consider the ultimate precursor of phillipsite to be glass shards derived from terrestrial volcanoes. Their proposal was based on a correlation between the distribution pattern of phillipsite in the Pacific and present-day global volcanicity.

Downhole Variability

Phillipsite is believed to form rapidly at the sediment/seawater interface and continue to grow within the sediment column (Bernat & Goldberg, 1969; Bernat *et al.*, 1970), although it starts to dissolve at depth (indicated by etched crystal faces and decreasing frequency of occurrence). At depth, phillipsite becomes increasingly rare and is rarely present at depths greater than 500 m (Kastner & Stonecipher, 1978). Clinoptilolites occur at greater depths than phillipsites and Iijima (1978) wonders whether phillipsite may alter to clinoptilolite in older or deeply buried sediments.

Indicator

Basaltic volcanism and perhaps low sedimentation rates.

Main Distinguishing Features

Elongate euhedral or subhedral prismatic crystals, common sector twinning, moderate to high relief. Low birefringence, association with altered volcanogenic material, especially palagonite. Phillipsite and clinoptilolite are not always easily distinguished in smear slides, although in co-occurrences, phillipsite usually occurs as the larger crystals.

NATIVE ELEMENTS

Native elements that can occur as important placer deposits include gold, platinum and diamond. Placers are metallic mineral deposits that have been formed by mechanical concentration. Most placers occur within a few kilometres of their source, thus most marine placers are restricted to the nearshore zone. Important gold and platinum placers occur off the coast of Alaska (Nelson & Hopkins, 1972; Cronan, 1980) and diamond placers have been located and mined off the coast of south-west Africa (Prentice, quoted in Cronan, 1980). Gold-bearing placers have also been reported off the coast of New South Wales, Australia (Jones & Davies, 1979). Gold usually occurs as rounded flattened grains which are golden yellow in reflected light and have a metallic lustre. Platinum most commonly occurs as worn, rough-textured flattened polygonal plates or irregular grains which are white to aluminium grey in reflected light (Milner, 1962). Both are, of course, opaque in transmitted light. Diamond occurs as clear, octahedral, often slightly rounded grains with very high relief. Grains often contain inclusions and are isotropic. It is possible that diamonds may be found long distances from their primary source due to their toughness and extreme hardness (Emery & Noakes, 1968). It is very unlikely that these native elements will be encountered in smear slides except in nearshore areas near known or suspected gold, platinum or diamond placers. A brief review of placer deposits is provided by Cronan (1980).

Few native elements occur in pelagic sediments. Cosmogenic nickel–iron will be described and very fine-grained carbon occurs derived from the burning of forests and grasslands. However, native copper has been reported from sedimentary sections at a number of DSDP sites. A vein of tiny copper crystals was cored 20 m above basalt on the eastern Atlantic lower continental rise at DSDP Site 105 (Hollister *et al.*, 1972). Donnelly & Nalli (1973) report the occurrence of native copper, occurring as aggregates of coarse octahedra (10–12 μm) on what appeared to be the inner surfaces of crude cylinders or irregular bodies a fraction of a millimetre in diameter. Consideration of probable redox potentials and the form of some of the aggregates led Donnelly & Nalli to tentatively conclude that the copper may have been concentrated and precipitated organically. The form of aggregates suggested the lining of worm tubes or the coating of faecal pellets. Blebs and strands

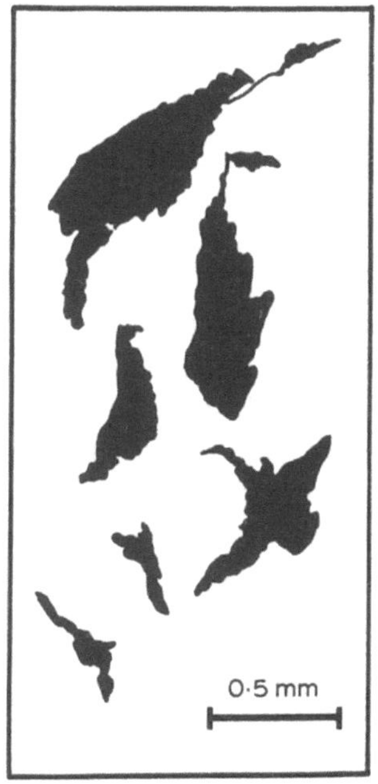

Fig. 116. Strands, wires and blebs of native copper, as they would appear in transmitted light. Lower Cretaceous, Manihiki Plateau, South Pacific (11°00.09′S, 162°15.78′W). DSDP Leg 33, 317A-23-3, 16–19 cm. (Drawn from Jenkyns, 1976).

of native copper up to 1 mm across were reported by Jenkyns (1976) in volcaniclastic sands up to 150 m above basaltic basement on the Manihiki Plateau in the central Pacific (Fig. 116). Jenkyns suggests that the copper was probably deposited from hydrothermal cupriferous solutions, derived from thermally driven seawater passing through fractured basalt (Bischoff, 1969; and others). Native copper often has a wirey and/or irregular form and will be opaque in transmitted light. However, in reflected light grains are copper-coloured. Grains will be malleable with a needle. Euhedral grains may, at first sight, resemble tarnished pyrite. However, chemical tests (e.g. copper turns nitric acid green) and X-ray diffraction will confirm the presence of copper.

EXTRATERRESTRIAL MATERIAL IN MARINE SEDIMENTS

Marine sediments are the repository for most of the extraterrestrial material accreted by the Earth (Brownlee, 1981). Cosmic material in marine sediments includes megascopic meteorites and meteoritic fragments, microscopic meteoritic particles and spherules and microscopic glassy particles called microtektites. Most cosmic particles are magnetic, due to indigenous metallic iron or to magnetite formation during atmospheric entry. They therefore can be separated from sediments by magnetic methods (Brownlee, 1981). Although extraterrestrial material forms a tiny fraction of most marine sediments, it is concentrated in pelagic sediments with low sedimentation rates, deposited far from land (such as in the central Pacific).

While megascopic meteorites are very rare in marine sediments, microscopic cosmic spherules can be relatively common, particularly in areas with low sedimentation rates. There are two main types of cosmic spherule—shiny, black, often conspicuous magnetic spheres and globules composed predominantly of magnetite, frequently with nickel–iron–cobalt metallic cores (called type I) and brown stony spherules consisting of olivine, glass and magnetite (called type S). A rarer type of sphere (called type G) is composed of magnetite, a small amount of glass and sometimes metal cores. Cosmic spherules range in size from 20 to 800 μm, although most are under 200 μm in diameter. However, not all spherules seen in marine sediments have a cosmic origin, for magnetic spherules can be produced by volcanic and industrial processes (Fredriksson & Martin, 1963; Brownlow *et al.*, 1965). Therefore it is best to consider only magnetic spherules with nickel–iron cores as being of certain extraterrestrial origin. Industrial spherules may be significant contaminants in sediment samples taken within a few centimetres of the ocean floor, especially offshore of industrialised regions or areas of the ocean floor receiving material from such localities. However, industrial contaminants are becoming increasingly widely distributed.

Extraterrestrial material appears to be especially concentrated within manganese nodules. Indeed, concentrations of nickel–iron spheres within nodules have been reported as being one to three orders of magnitude higher than concentrations in sediment (Finkelman, 1970; 1972). Manganese nodules accrete very slowly and dense particles such as cosmic spherules settle and remain in cracks and crevasses on the nodule while the less dense sedimentary material is mostly washed off and dispersed by weak bottom currents. As manganese nodules grow so slowly and remain on the sediment surface, cosmic spherules within a manganese nodule may represent cosmic material that fell upon the nodule over some millions of years (Brownlee, 1981).

Estimates have been made of the influx rates of extraterrestrial material based on spherule concentrations in sedimentary deposits. However, these vary by many orders of magnitude but most estimates are around 10^4–10^5 tons per year (Schmidt, 1965; Ivanov & Florenskiy, 1970).

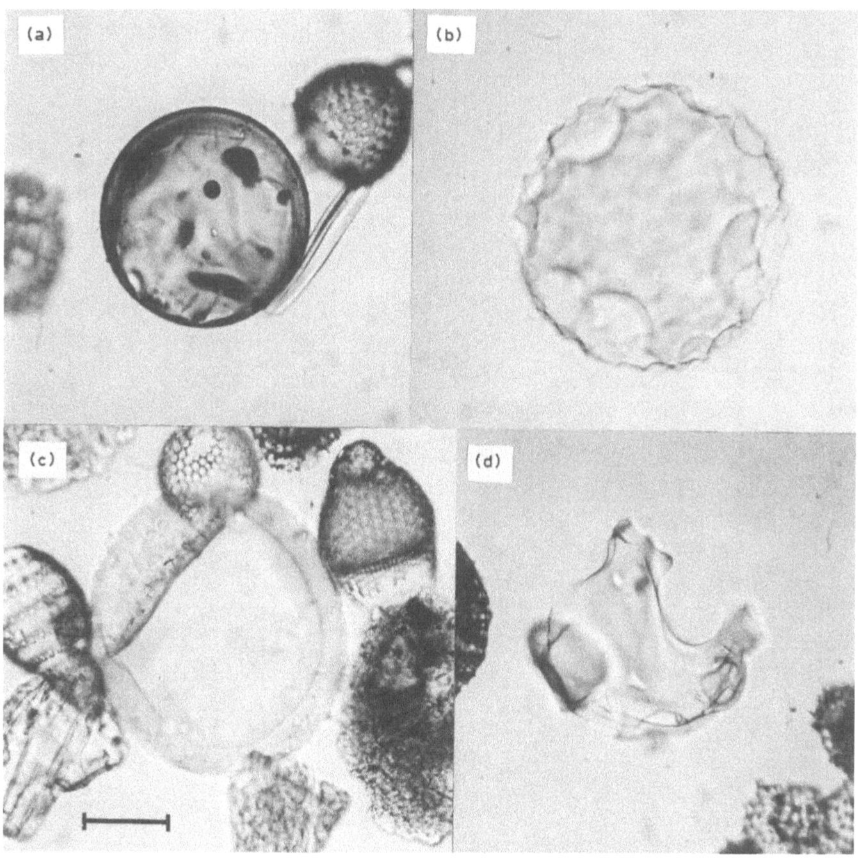

Fig. 117. Late Eocene microtektites from onshore localities on Barbados. In (a) and (c), radiolarian tests are also seen in the field of view. Plane-polarised light, scale bar equals 0·05 mm. (Courtesy of Dr A. Sanfilippo).

In smear slides, metallic cosmic spherules will appear opaque and may resemble spheroidal micronodules. They may be more common in samples taken from areas of low sedimentation. The geologist should always bear in mind that not all spherules are of cosmic origin. Criteria for identifying true cosmic spherules are discussed by Wright & Hodge (1965), Finkelmann (1972), Del Monte *et al.* (1975) and Blanchard & Davis (1978). These rely on the determination of elemental abundances within the spherule, using specialist analytical techniques, and comparison with cosmic and terrestrial abundances. Presence of iron–nickel

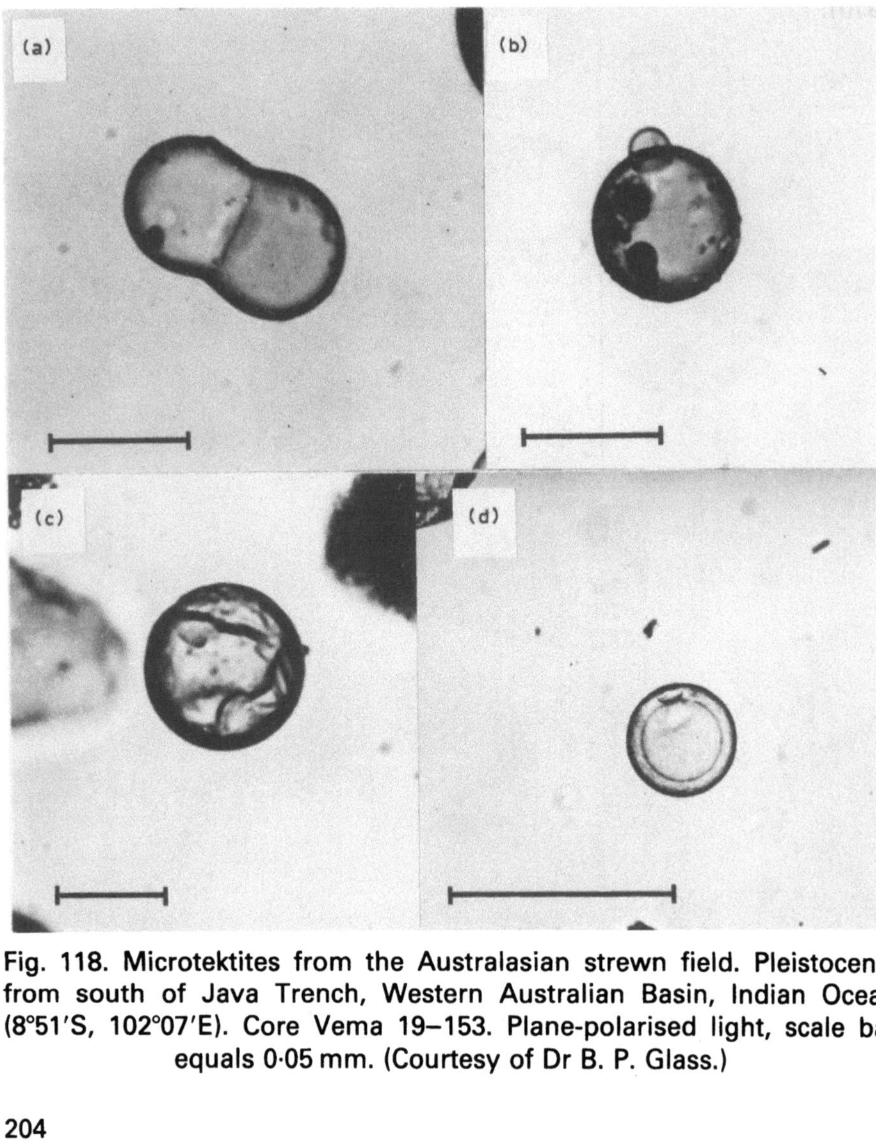

Fig. 118. Microtektites from the Australasian strewn field. Pleistocene, from south of Java Trench, Western Australian Basin, Indian Ocean (8°51′S, 102°07′E). Core Vema 19–153. Plane-polarised light, scale bar equals 0·05 mm. (Courtesy of Dr B. P. Glass.)

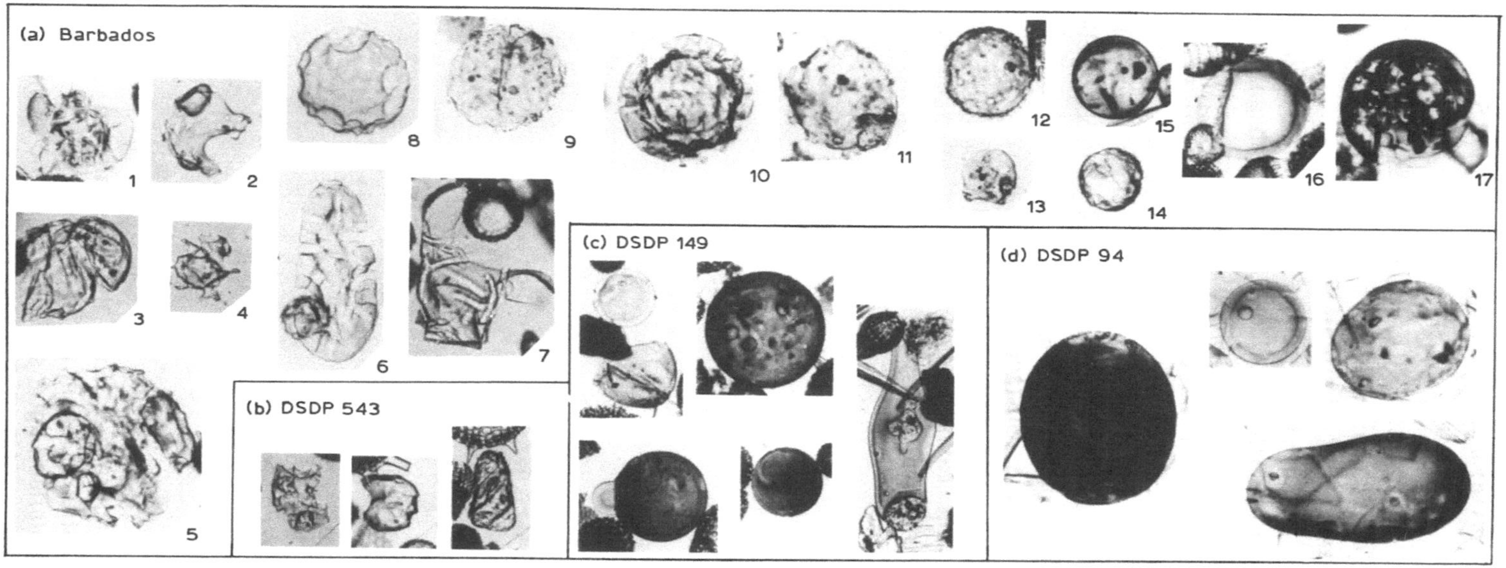

Fig. 119. Examples of microtektites from: (a) Barbados (onshore localities); (b) DSDP Site 543 (seaward of Barbados Ridge Complex); (c) DSDP Site 149 (central Venezuelan Basin); and (d) DSDP Site 94 (continental slope, Yucatan Platform). All Late Eocene and from the North American strewn field. Original magnification × 105. (Courtesy of Dr A. Sanfilippo. Reprinted with permission from *Nature,* **314** (6012), pp. 613–15. Copyright © 1985, Macmillan Journals Ltd.)

cores with meteoritic composition (Fredriksson & Martin, 1963) is considered indicative of a cosmic origin.

Microscopic glassy particles (<1 mm in diameter) may also be seen in deep-sea sediments, especially those recovered from the Indian Ocean, Philippine Sea, equatorial Atlantic Ocean and Caribbean Sea. These have been identified as microtektites on consideration of their age (from fission track dating), chemical composition and geographical distribution. Most microtektites are spherical but they can be oval, dumb-bell, teardrop, discoidal or irregular in shape (Figs 117–122). They may vary from colourless through shades of green, yellow and brown to black. They are generally transparent, although a few may be translucent or even opaque. Surface texture may vary from smooth and glassy to irregular and grooved and pitted. They frequently contain spherical bubble cavities (Fig. 122) and may be corroded. Microtektites consist of silicate glass and have refractive indices within

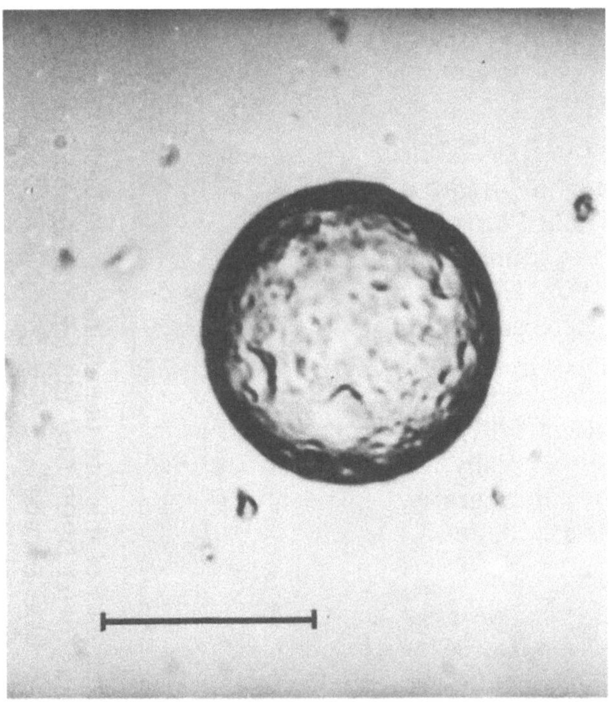

Fig. 120. Spherical microtektite from the Australasian strewn field. Pleistocene, South Australian Basin (41°21.8′S, 114°08.0′E). Core Robert Conrad 9-143. Plane-polarised light, scale bar equals 0·05 mm. (Courtesy of Dr B. P. Glass.)

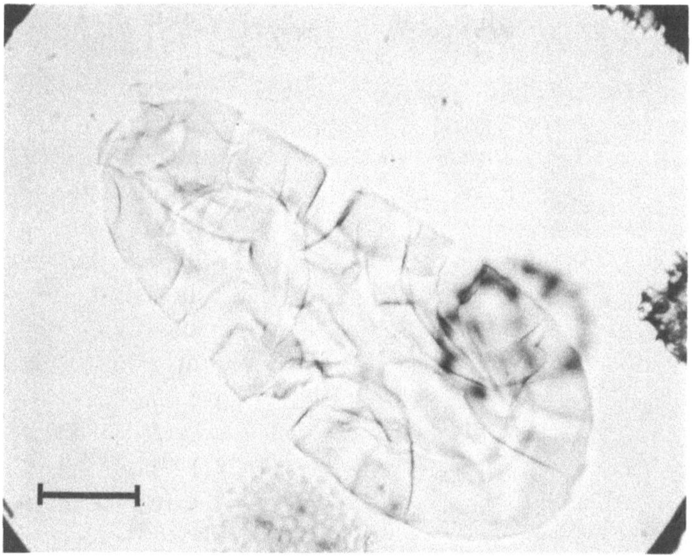

Fig. 121. Teardrop-shaped microtektite with marginal embayments. Late Eocene, Barbados. Plane-polarised light, scale bar equals 0·05 mm. (Courtesy of Dr A. Sanfilippo.)

the range 1·46 to 1·62 (Glass, 1968). Although there has been much debate on tektite origin in the past, it is now thought that they represent solidified terrestrial impact melt ejected during large meteorite or cometary impacts (Faul, 1966; Urey, 1971; and others). The theory that they owe their origin to lunar impacts or even lunar volcanoes (Chapman & Larson, 1963; O'Keefe, 1970; Chapman, 1971; and others), is now not widely held, as tektite elemental compositions are very similar to terrestrial sedimentary rocks and unlike known lunar rocks (Brownlee, 1981). Tektites occur in widely strewn fields and as they record geologically instantaneous events, they have potential for excellent chronostratigraphic correlation (Glass *et al.*, 1979). Within ocean sediments, three distinct microtektite occurrences have been found:

(1) A 0·7 million year old layer in the Indian Ocean, Philippine Sea and western equatorial Pacific. This forms part of the Indo–China/Australasian strewn field (Glass, 1972; Glass *et al.*, 1979).

(2) A Late Eocene (34 million year old) layer found in the Caribbean Sea, Gulf of Mexico and onshore Barbados (Saunders *et al.*, 1984). This is considered part of the North American strewn field (Glass *et al.*, 1973).

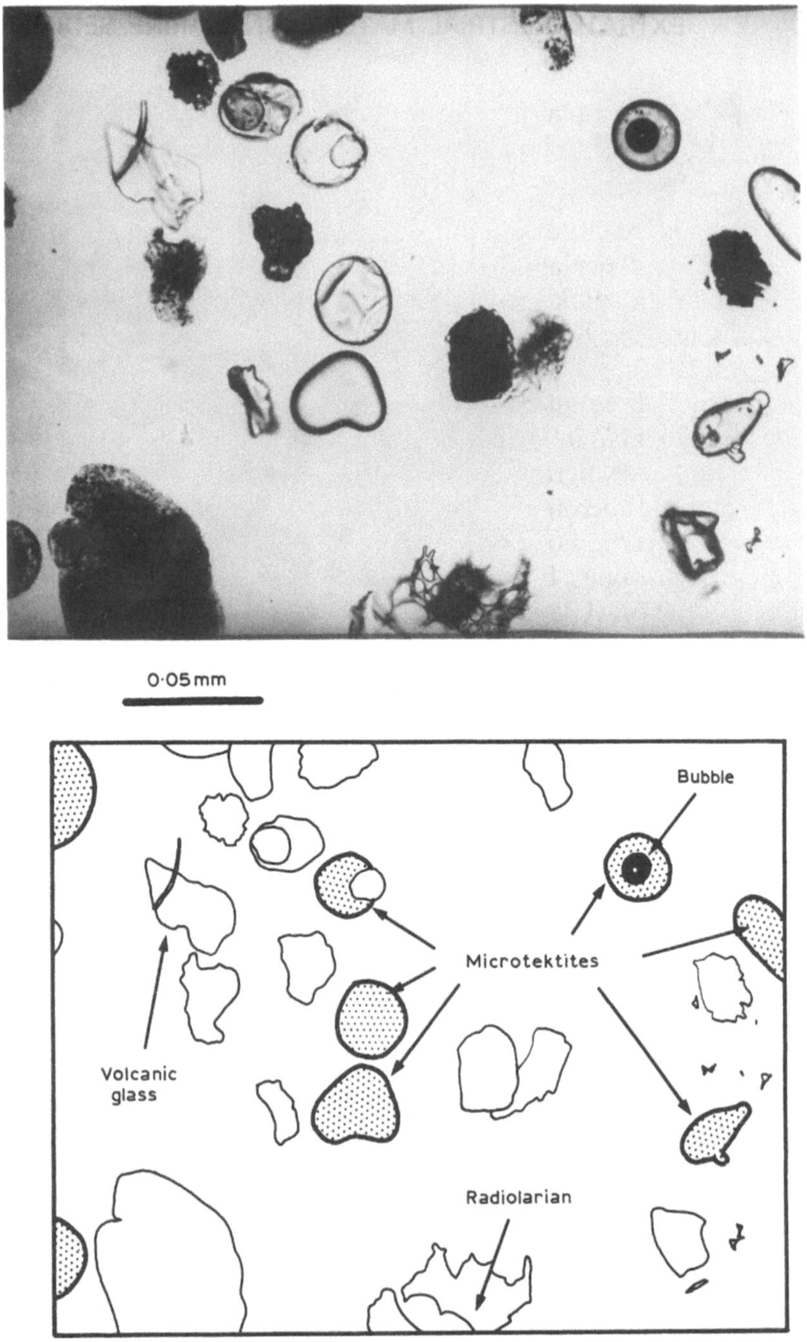

Fig. 122. Field of view showing a number of microtektites with line drawing beneath. Volcanic glass and radiolarians are also present in the field. Pleistocene, from south of Java Trench, Western Australian Basin, Indian Ocean (8°51′S, 102°07′E). Core Vema 19-153. Plane-polarised light. (Courtesy of Dr B. P. Glass.)

(3) A layer in the equatorial Atlantic, approximately 1 million years old. This is considered to be part of the Ivory Coast strewn field (Glass, 1968).

The colour of microtektites, as seen in transmitted light, seems characteristic for particular strewn fields. The Ivory Coast microtektites are olive-green or dark brown in colour, while the Australasian microtektites are mostly yellowish-brown or colourless to yellowish-green (Glass, 1968).

Although only three microtektite layers have so far been recognised in marine sediments, it is considered likely that other microtektite layers exist in the stratigraphic column (Glass & Perch-Nielsen, 1986). Discovery of new microtektite occurrences would be of considerable geological importance. For not only would they reveal hitherto unrecognised impact events but would provide excellent potential for improved correlation (Glass & Perch-Nielsen, 1986).

REFERENCES

ALEVA, G. J. J. (1973). Aspects of the historical and physical geology of the Sunda shelf essential to the exploration of submarine tin placers. *Geologie En Mijnbouw,* **52,** 79–91.

ALLEN, J. E. (1956). Estimation of percentages in thin sections—considerations of visual psychology. *Journal of Sedimentary Petrology,* **26,** 160–1.

ALLMAN, M. & LAWRENCE, D. F. (1972). *Geological Laboratory Techniques.* Blandford Press, London.

ANDEL, Tj. H. VAN, (1975). Mesozoic/Cenozic calcite compensation depth and the global distribution of calcareous sediments. *Earth and Planetary Science Letters,* **26,** 187–94.

ANDEL, Tj. H. VAN, HEATH, G. R. & MOORE, T. C. (1975). Cenozoic history and paleoceanography of the central equatorial Pacific Ocean: a regional synthesis of Deep Sea Drilling Project data. *Geological Society of America Memoir,* **143,** 134 pp.

ARRHENIUS, G. (1963). Pelagic sediments. In *The Sea, Vol. 3,* ed. M. N. Hill. John Wiley, New York, pp. 655–727.

ARRHENIUS, G. & BONATTI, E. (1965). Neptunism and volcanism in the ocean. *Progress in Oceanography,* **3,** 7–22.

BAILEY, E. H. & STEVENS, R. E. (1960). Selective staining of K-feldspar and plagioclase on rock slabs and thin sections. *American Mineralogist,* **45,** 1020–5.

BASS, M. N., MOBERLY, R., RHODES, J. M., SHIH CHI-YU & CHURCH, S. E. (1973). Volcanic rock cores in the central Pacific, Leg 17, Deep Sea Drilling Project. *Initial Reports of the Deep Sea Drilling Project,* **17,** 429–503.

BASU, A. (1985). Reading provenance from detrital quartz. In *Provenance of Arenites,* ed. G. G. Zuffa, D. Reidel, Dordrecht, pp. 231–47.

BEHRENS, E. W. & LAND, L. S. (1972). Subtidal Holocene dolomite, Baffin Bay, Texas. *Journal of Sedimentary Petrology,* **42,** 155–61.

BERGER, W. H. (1968). Planktonic foraminifera: selective solution and palaeoclimatic interpretation. *Deep Sea Research,* **15,** 31–43.

BERGER, W. H. (1976). Biogenous deep-sea sediments: production, preservation and interpretation. In *Chemical Oceanography, Vol. 5,* ed. J. P. Riley & R. Chester. Academic Press, New York, pp. 265–338.

BERGER, W. H. & VON RAD, U. (1972). Cretaceous and Cenozoic

sediments from the Atlantic Ocean. *Initial Reports of the Deep Sea Drilling Project,* **14,** 787–954.

BERNAT, M. & GOLDBERG, E. D. (1969). Thorium isotopes in the marine environment. *Earth and Planetary Science Letters,* **5,** 308–12.

BERNAT, M., BIERI, R. H., KOIDE, M., GRIFFIN, J. J. & GOLDBERG, E. D. (1970). Uranium, thorium, potassium and argon in marine phillipsites. *Geochimica et Cosmochimica Acta,* **34,** 1053–71.

BERNER, R. A. (1969). The synthesis of framboidal pyrite. *Economic Geology,* **64,** 383–4.

BERNER, R. A. (1970). Sedimentary pyrite formation. *American Journal of Science,* **268,** 1–23.

BERNER, R. A. (1984). Sedimentary pyrite formation: an update. *Geochimica et Cosmochimica Acta,* **48,** 605–15.

BISCAYE, P. E. (1965). Mineralogy and sedimentation of Recent deep-sea clay in the Atlantic Ocean and adjacent seas and oceans. *Bulletin of the Geological Society of America,* **76,** 803–32.

BISCHOFF, J. L. (1969). Red Sea geothermal brine deposits: their mineralogy, chemistry and genesis. In *Hot Brines and Recent Heavy Metal Deposits in the Red Sea,* ed. E. T. Degens & D. A. Ross. Springer-Verlag, New York, pp. 368–401.

BLANCHARD, M. B. & DAVIS, A. (1978). Analysis of ablation debris from natural and artificial iron meteorites. *Journal of Geophysical Research,* **83,** 1793–808.

BOLES, J. R. & WISE, W. S. (1978). Nature and origin of deep-sea clinoptilolite. In *Natural Zeolites,* eds L. B. Sand & F. A. Mumpton. Pergamon Press, Oxford, pp. 235–43.

BONATTI, E. (1963). Zeolites in Pacific pelagic sediments. *Transactions of the New York Academy of Sciences,* **25,** 938–48.

BONATTI, E. (1965). Palagonite, hyaloclastites and alteration of volcanic glass in the ocean. *Bulletin Volcanologique,* **28,** 1–15.

BONATTI, E. (1966). Deep sea authigenic calcite and dolomite. *Science,* **153,** 534–6.

BONATTI, E. (1968). Mechanisms of deep sea volcanism in the south Pacific. In *Researches in Geochemistry,* ed. P. H. Abelson. John Wiley, New York, pp. 453–91.

BONATTI, E. (1981). Metal deposits in the oceanic lithosphere. In *The Sea, Vol. 7, The Oceanic Lithosphere,* ed. C. Emiliani. Wiley Interscience, New York, pp. 639–86.

BONATTI, E. & NAYUDA, Y. (1965). The origin of manganese nodules on the ocean floor. *American Journal of Science,* **263,** 17–39.

BONTE, P., LALOU, C. & LATOUCHE, C. (1980). Ferromanganese deposits in cores from the Kane and Atlantis fracture areas: possible relationships with hydrothermalism. *Journal of the Geological Society,* **137,** 373–7.

BORCHERT, H. & BRAUN, H. (1963). Zum chemismus von drei glaukonittypen. *Chemie der Erde,* **23,** 82–90.

BORELLA, P. & ADELSECK, C. (1979). Manganese micronodules in sediments: a subsurface in-situ origin, Leg 51, Deep Sea Drilling Project. *Initial Reports of the Deep Sea Drilling Project,* **51/53,** 771–87.

BOSTROM, K., FARQUARSON, B. & EYL, W. (1971). Submarine hot springs as a source of active ridge sediments. *Chemical Geology,* **10,** 189–203.

BOWEN, H. J. M. (1956). Strontium and barium in seawater and marine organisms. *Journal Marine Biological Association UK,* **35,** 451–60.

BRAMLETTE, M. N. (1946). The Monterey Formation of California and the origin of siliceous rocks. *US Geological Survey Professional Paper,* **212,** 55 pp.

BRAMLETTE, M. N. (1961). Pelagic sediments. In *Oceanography,* ed. M. Sears. American Association for the Advancement of Science (Publication 67), Washington DC, pp. 345–66.

BRISKIN, M. & SCHREIBER, B. C. (1978). Authigenic gypsum in marine sediments. *Marine Geology,* **28,** 37–49.

BROWNLEE, D. E. (1981). Extraterrestrial components. In *The Sea, Vol. 7, The Oceanic Lithosphere.* ed. C. Emiliani. Wiley Interscience, New York, pp. 733–62.

BROWNLOW, A. E., HUNTER, W. & PARKIN, D. W. (1965). Cosmic dust collections at various latitudes. *Geophysical Journal of the Royal Astronomical Society,* **9,** 337–88.

BURST, J. F. (1958*a*). "Glauconite" pellets: their mineral nature and application to stratigraphic interpretations. *Bulletin of the American Association of Petroleum Geologists,* **42,** 310–27.

BURST, J. F. (1958*b*). Mineral heterogeneity in 'glauconite' pellets. *American Mineralogist,* **43,** 481–97.

BURT, F. A. (1932). Formative processes in concretions formed about fossils as nuclei. *Journal of Sedimentary Petrology,* **2,** 38–45.

CARVER, R. E. (1971). Heavy-mineral separation. In *Procedures in Sedimentary Petrology,* ed. R. E. Carver. Wiley Interscience, New York, pp. 427–52.

CHAMLEY, H. (1967). Possibilities d'utilisation de la cristallinite d'un mineral argileux (illite) comme temoin climatique dans les sediments recents. *Compte rendu de l'Académie des Sciences,* **265D,** 184–7.

CHAPMAN, D. R. (1971). Australasian tektite geographic pattern, crater and ray of origin and theory of tektite events. *Journal of Geophysical Research,* **76,** 6309–38.

CHAPMAN, D. R. & LARSON, H. K. (1963). On the lunar origin of tektites. *Journal of Geophysical Research,* **68,** 4305–58.

CHEMLIK, F. B. (1967). Electro-osmotic core cutting. *Marine Geology,* **5,** 321–5.

CHESSELET, R., JEDWAB, J., DARCOURT, C. & DEHAIRS, F. (1976). Barite as discrete suspended particles in the Atlantic Ocean. *Transactions of the American Geophysical Union*, **57**, 255 (abs.).

CHURCH, T. M. (1970). Marine barite. PhD thesis, University of California, San Diego, 133 pp (unpublished).

CHURCH, T. M. & BERNAT, M. (1972). Thorium and uranium in marine barite. *Earth and Planetary Science Letters*, **14**, 139–44.

CLOUD, P. E. (1955). Physical limits of glauconite formation. *Bulletin of the American Association of Petroleum Geologists*, **39**, 484–92.

COOK, H. E. & ZEMMELS, I. (1972). X-ray mineralogy studies—Leg 9. *Initial Reports of the Deep Sea Drilling Project*, **9**, 707–78.

CRIDDLE, A. J. (1974). A preliminary description of microcrystalline pyrite from the nannoplankton ooze at Site 251, southwest Indian Ocean. *Initial Reports of the Deep Sea Drilling Project*, **26**, 603–7.

CRONAN, D. S. (1974). Authigenic minerals in deep-sea sediments. In *The Sea, Vol. 5*, ed. E. D. Goldberg. Wiley-Interscience, New York, pp. 491–525.

CRONAN, D. S. (1980). *Underwater Minerals*. Academic Press, London.

DALRYMPLE, G. B. & LAMPHERE, M. A. (1970). *Potassium–Argon Dating*. W. H. Freeman, London.

DAMUTH, J. E. & FAIRBRIDGE, R. W. (1970). Equatorial Atlantic deep-sea arkosic sands and ice-age aridity in tropical South America. *Geological Society of America Bulletin*, **81**, 189–206.

DAPPLES, E. C. (1967). Diagenesis of sandstones. In *Diagenesis in Sediments*, eds G. Larsen & G. V. Chilingar. Elsevier, Amsterdam, pp. 91–125.

DAVIES, T. A. & GORSLINE, D. S. (1976). Oceanic sediments and sedimentary processes. In *Chemical Oceanography, Vol. 5*, eds J. P. Riley & P. Chester. Academic Press, New York, pp. 1–80.

DEAN, W. E. & SCHREIBER, B. C. (1977). Authigenic barite, Leg 41 Deep Sea Drilling Project. *Initial Reports of the Deep Sea Drilling Project*, **41**, 915–31.

DEAN, W. E., LEINEN, M. & STOW, D. A. V. (1985). Classification of deep-sea fine-grained sediments. *Journal of Sedimentary Petrology*, **55**, 250–6.

DEL MONTE, M., NANNI, T. & TAGLIAZUCCA, M. (1975). Ferromagnetic volcanic particulate matter and black magnetic spherules: a comparative study. *Journal of Geophysical Research*, **80**, 1880–4.

DESAI, M. V. M., KOSHY, E. & GANGULY, A. K. (1969). Solubility of barium in seawater in the presence of dissolved organic matter. *Current Science*, **38**, 107–8.

DIESTER-HAASS, L. (1976). Late Quaternary climatic variations in Northwest Africa deduced from east Atlantic sediment cores. *Quaternary Research*, **6**, 299–314.

DIETRICH, G., KALLE, K., KRAUS, W. & SIEDLER, G. (1980). *General Oceanography (2nd Edition).* Wiley Interscience, New York.

DONNELLY, T. W. & NALLI, G. (1973). Mineralogy and chemistry of Caribbean sediments. *Initial Reports of the Deep Sea Drilling Project,* **15,** 929–61.

DOYLE, L. J., CLEARY, W. J. & PILKEY, O. H. (1968). Mica: its use in determining shelf depositional regimes. *Marine Geology,* **6,** 381–9.

DUNHAM, R. (1962). Classification of carbonate rocks according to depositional texture. In *Classification of Carbonate Rocks,* ed. W. E. Ham. American Association of Petroleum Geologists, Tulsa, pp. 108–21.

EDELMAN, C. H. & DOEGLAS, D. J. (1932). Relikstrurkturen detritischer Pyroxenen und Amphibolen. *Tschermak's Mineralogische u. Petrographische Mittheilungen,* **42,** 482–9.

ELDERFIELD, H. (1976). Hydrogenous material in marine sediments excluding manganese nodules. In *Chemical Oceanography, Vol. 5,* eds J. P. Riley & R. Chester. Academic Press, London, pp. 137–215.

EMERY, K. O. & NOAKES, L. C. (1968). Economic placer deposits of the continental shelf. *Economic Commission for Asia and the Far East: Committee for Co-ordinating of Joint Prospecting (CCOP) Technical Bulletin,* **1,** 95–111.

FAIRBRIDGE, R. W. (1967). Phases of diagenesis and authigenesis. In *Diagenesis in Sediments,* eds G. Larsen & G. V. Chilingar. Elsevier, Amsterdam. pp. 19–89.

FAUL, H. (1966). Tektites are terrestrial. *Science,* **152,** 1341–5.

FINKELMAN, R. B. (1970). Magnetic particles extracted from manganese nodules: suggested origin from stony and iron meteorites. *Science,* **167,** 982–4.

FINKELMAN, R. B. (1972). Relationship between manganese nodules and cosmic spherules. *Marine Technology Society Journal,* **6,** 34–9.

FISCHER, A. G., HONJO, S. & GARRISON, R. E. (1967). *Electron Micrographs of Limestones and their Nannofossils.* Princeton University Press, Princeton, New Jersey.

FISHER, R. V. (1964). Maximum size, median diameter and sorting of tephra. *Journal of Geophysical Research,* **69,** 341–57.

FISHER, R. V. (1965). Settling velocity of glass shards. *Deep Sea Research,* **12,** 345–53.

FOLK, R. L. (1951). A comparison chart for visual percentage estimation. *Journal of Sedimentary Petrology,* **21,** 32–3.

FOLK, R. L. (1980). *Petrology of Sedimentary Rocks.* Hemphill, Austin, Texas.

FRANCIS, T. J. G. (1986). The development and transfer of technology in scientific ocean drilling. *Proceedings of Greenwich Forum XII, Offshore and Maritime Technology Transfer Conference.* Spearhead, Kingston-upon-Thames.

FREDRIKSSON, K. & MARTIN, L. R. (1963). The origin of black spherules found in Pacific islands, deep sea sediments and Antarctic ice. *Geochimica et Cosmochimica Acta,* **27,** 245–8.

FRESNEL, J., GALLE, P. & GAYRAL, P. (1979). Résultats de la microanalyse des cristaux vacuolaires chez deux Chromatophytes unicellulaires marines: *Exanthemachrysis gayraliae, Pavlova* sp. (Prymnésiophycées, Pavlovacées). *Compte Rendu hebdomadaire des Séances de l'Académie des Sciences,* **288D,** 823–5.

FURUTA, T. & ARAI, F. (1980). Petrographic and geochemical properties of tephras in Deep Sea Drilling Project cores from the North Philippine Sea. *Initial Reports of the Deep Sea Drilling Project,* **58,** 617–25.

GARRELS, R. M. & MACKENZIE, F. T. (1971). *Evolution of Sedimentary Rocks.* Norton, New York.

GEORGE, W. O. (1924). The relation of the physical properties of natural glasses to their chemical composition. *Journal of Geology,* **32,** 353–72.

GLASBY, G. P. (ed.) (1977). *Marine Manganese Deposits.* Elsevier, Amsterdam (Elsevier Oceanography Series 15).

GLASS, B. P. (1968). Glassy objects (microtektites?) from deep sea sediments near Ivory Coast. *Science,* **161,** 891–3.

GLASS, B. P. (1972). Australasian microtektites in deep sea sediments. *Antarctic Research Series,* **19,** 335–48.

GLASS, B. P. & PERCH-NIELSEN, K. (1986). Microtektites in micropaleontological samples. *Micropaleontology,* **32,** 46–7.

GLASS, B. P., BAKER, R. N., STORZER, D., & WAGNER, G. A. (1973). North American microtektites from the Caribbean Sea and their fission track age. *Earth and Planetary Science Letters,* **19,** 184–92.

GLASS, B. P., SWINCKI, M. B. & ZWART, P. A. (1979). Australasian, Ivory Coast and North American tektite strewn fields: size, mass and correlation with geomagnetic reversals and other earth events. *Proceedings of the Tenth Lunar and Planetary Science Conference,* Pergamon, New York, pp. 2535–45.

GOLDBERG, E. D. (1954). Chemical scavengers of the sea. *Journal of Geology,* **62,** 249–65.

GOLDBERG, E. D. & GRIFFIN, J. J. (1964). Sedimentation rates and mineralogy in the South Atlantic. *Journal of Geophysical Research,* **69,** 4293–309.

GOLDBERG, E. D. & GRIFFIN, J. J. (1970). The sediments of the northern Indian Ocean. *Deep Sea Research,* **17,** 513–37.

GOLDBERG, E. D., SOMAYAJULU, B. L. K. & GALLOWAY, J. (1969). Differences between barites of marine and continental origins. *Geochimica et Cosmochimica Acta,* **33,** 287–9.

GOLDICH, S. S. (1938). A study in rock weathering. *Journal of Geology,* **46,** 17–58.

GOODAY, A. J. & NOTT, J. A. (1982). Intracellular barite crystals in two xenophyophores, *Aschemonella ramuliformis* and *Galatheammina* sp. (Protozoa : Rhizopoda) with comments on the taxonomy of *A. ramuliformis*. *Journal Marine Biological Association of UK,* **62,** 595–605.

GRIFFITHS, J. C. & ROSENFELD, M. A. (1954). Operator variation in experimental research. *Journal of Geology,* **62,** 74–91.

HAQ, B. U. & BOERSMA, A. (eds) (1978). *Introduction to Marine Micropalaeontology*. Elsevier, New York.

HAWKINS, D. B. (1961). Experimental hydrothermal studies bearing on rock weathering and clay mineral formation. PhD thesis, Pennsylvania State University (unpublished).

HAY, R. L. (1966). Zeolites and zeolitic reactions in sedimentary rocks. *Geological Society of America Special Paper,* **85,** 130 pp.

HAYES, J. R. & KLUGMAN, M. A. (1959). Feldspar staining methods. *Journal of Sedimentary Petrology,* **29,** 227–32.

HEATH, G. R. (1969). Mineralogy of Cenozoic deep-sea sediments from the equatorial Pacific Ocean. *Geological Society of America Bulletin,* **80,** 1997–2018.

HEATH, G. R. (1982). Deep-sea ferromanganese nodules. In *The Environment of the Deep Sea,* eds W. G. Ernst & J. G. Morin. Prentice-Hall, Englewood Cliffs, New Jersey, pp. 105–53.

HEATH, G. R. & MOBERLY, R., (1971). Cherts from the western Pacific, Leg 7, Deep Sea Drilling Project. *Initial Reports of The Deep Sea Drilling Project,* **7,** 991–1007.

HERMAN, Y. (1978). Pteropods. In *Introduction to Marine Micropalaeontology,* eds B. U. Haq & A. Boersma. Elsevier, New York, pp. 151–9.

HERZOG, L. F., PINSON, W. H. & CORMIER, R. F. (1958). Sediment age determination by Rb/Sr analysis of glauconite. *Bulletin of the American Association of Petroleum Geologists,* **42,** 717–33.

HOFFERT, M., PERSON, A., COURTOIS, C., KARPOFF, A. M. & TRAUTH, D. (1980). Sedimentology, mineralogy and geochemistry of hydrothermal deposits from Holes 424, 424A, 424B and 424C (Galapagos Spreading centre). *Initial Reports of the Deep Sea Drilling Project,* **54,** 339–76.

HOLLISTER, C. D. *et al.* (1972). *Initial Reports of the Deep Sea Drilling Project, Vol. 11.* US Government Printing Office, Washington DC.

HOLMES, M. A., BREZA, J. R. & WISE, S. W. (1987). Provenance and deposition of Lower Cretaceous turbidite sands at Deep Sea Drilling Project Site 603, lower continental rise, North Carolina. *Initial Reports of the Deep Sea Drilling Project,* **93**(2)**,** 941–60.

HONNOREZ, J. (1978). Generation of phillipsites by palagonitisation of basaltic glass in seawater and the origin of K-rich deep sea sediments. In *Natural Zeolites,* eds L. B. Sand & F. A. Mumpton. Pergamon Press, Oxford, pp. 245–58.

HUANG, T. C., WATKINS, N. D. & SHAW, D. M. (1975), Atmospherically transported volcanic glass in deep-sea sediments: development of a separation and counting technique. *Deep Sea Research,* **22,** 185–96.
HUBERT, G., RIEDER, N., SCHMITT, G. & SEND, W. (1975). Bariumanreicherung in den Müllerschen Körperchen der Loxodidae (Ciliata, Holotricha). *Zeitschrift für Naturforschung,* **30C,** p. 422.
HUBERT, J. F. (1962). A zircon–tourmaline–rutile maturity index and the interdependence of the composition of heavy mineral assemblages with the gross composition and texture of sandstones. *Journal of Sedimentary Petrology,* **32,** 440–50.
HUBERT, J. F. (1971). Analysis of heavy mineral assemblages. In *Procedures in Sedimentary Petrology,* ed. R. E. Carver. Wiley Interscience, New York, pp. 453-78.
HURLEY, P. M., CORMIER, R. F., HOWER, J., FAIRBAIRN, H. W. & PINSON, W. H. (1960). Reliability of glauconite for age measurement by K–Ar and Rb–Sr methods. *Bulletin of the American Association of Petroleum Geologists,* **44,** 1793–808.
IIJIMA, A. (1978). Geological occurrences of zeolites in marine environments. In *Natural Zeolites,* ed. L. B. Sand & F. A. Mumpton. Pergamon Press, Oxford, pp. 175–98.
IVANOV, A. V. & FLORENSKIY, K. P. (1970). The rate of fall of cosmic dust on the Earth. *Geokhimiya,* **11,** 1365–72 (in Russian).
JENKYNS, H. C. (1976). Sediments and sedimentary history of the Manihiki Plateau, south Pacific Ocean. *Initial Reports of the Deep Sea Drilling Project,* **33,** 873–90.
JIPA, D. & KIDD, R. B. (1974). Sedimentation of coarser-grained interbeds in the Arabian Sea and sedimentation processes of the Indus Cone. *Initial Reports of the Deep Sea Drilling Project,* **23,** 471–95.
JONES, E. J. W. (1973). Volcanic glass in abyssal clays sampled at DSDP Leg 20 drilling sites, northwest Pacific. *Initial Reports of the Deep Sea Drilling Project,* **20,** 389–402.
JONES, H. A. & DAVIES, P. J. (1979). Preliminary studies of offshore placer deposits, Eastern Australia. *Marine Geology,* **30,** 243–68.
JONES, J. B. & SEGNIT, E. R. (1971). The nature of opal, I. Nomenclature and constituent phases. *Journal of the Geological Society of Australia,* **18,** 56–68.
KASTNER, M. (1981). Authigenic silicates in deep-sea sediments: formation and diagenesis. In *The Sea, Vol. 7, The Oceanic Lithosphere,* ed. C. Emiliani. Wiley Interscience, New York, pp. 915–80.
KASTNER, M. & GIESKES, J. M. (1976). Interstitial water profiles and sites of diagenetic reactions, Leg 35, DSDP, Bellingshausen Abyssal Plain. *Earth and Planetary Science Letters,* **33,** 11–20.
KASTNER, M. & SIEVER, R. (1979). Low temperature feldspars in sedimentary rocks. *American Journal of Science,* **279,** 435–79.

KASTNER, M. & STONECIPHER, S. A. (1978). Zeolites in pelagic sediments of the Atlantic, Pacific and Indian Oceans. In *Natural Zeolites,* eds L. B. Sand & F. A. Mumpton. Pergamon Press, Oxford, pp. 199–220.

KELLER, W. D. (1953). Illite and montmorillonite in green sedimentary rocks. *Journal of Sedimentary Petrology,* **23,** 3–9.

KELLER, W. D. (1958). Glauconitic mica in the Morrison Formation in Colorado. *Clays and Clay Minerals,* **5,** 120–8.

KELTS, K. & McKENZIE, J. A. (1976). Cretaceous volcanogenic sediments from the Line Island chain: diagenesis and formation of K-feldspar, DSDP Leg 33, Hole 315A and site 316. *Initial Reports of the Deep Sea Drilling Project,* **33,** 789–803.

KERR, P. F. (1977). *Optical Mineralogy (4th Edition).* McGraw-Hill, New York.

KHARIN, G. S. (1974). Chemigenic aragonite on the seafloor in some abyssal areas of the Atlantic Ocean. *Oceanology,* **14,** 232–7.

KIDD, R. B. (1975). Holocene sedimentation in the Tyrrhenian Basin of the Mediterranean Sea. PhD thesis, University of Southampton (unpublished).

KIDD, R. B., SEARLE, R. C., RAMSAY, A. T. S., PRICHARD, H. & MITCHELL, J. (1982). The geology and formation of Kings Trough, Northeast Atlantic Ocean. *Marine Geology,* **48,** 1–30.

KOLLA, V. & BISCAYE, P. E. (1973). Deep-sea zeolites: variations in space and time in sediments of the Indian Ocean. *Marine Geology,* **15,** 11–17.

KOLLA, V., BISCAYE, P. E. & HANLEY, A. F. (1979). Distribution of quartz in Late Quaternary Atlantic sediments in relation to climate. *Quaternary Research,* **11,** 261–77.

KOLODNY, Y. (1981) Phosphorites. In *The Sea, Vol. 7, The Oceanic Lithosphere,* ed. C. Emiliani. Wiley Interscience, New York, pp. 981–1023.

KRAUS, E. H., HUNT, W. F. & RAMSDELL, L. S. (1959). *Mineralogy.* McGraw-Hill, New York.

KREGER, D. R. & BOERÉ, H. (1969). Some observations on barium sulphate in *Spirogyra. Acta Botanica Neerlandica,* **18,** 143–51.

KRYNINE, P. D. (1946). Microscopic morphology of quartz types. *Proceedings of the 2nd Pan American Congress of Mineralogists, Engineers and Geologists, 3, 2nd Comm.,* 35–49. (See Folk, 1980.)

KU, T. L., (1977). Rates of accretion. In *Marine Manganese Deposits,* ed. G. P. Glasby. Elsevier, Amsterdam (Elsevier Oceanography Series 15), pp. 249–67.

LALOU, C., BRICHET, E. & BONTE, P. (1980). Some new data on the genesis of manganese nodules. In *Geology and Geochemistry of Manganese,* Vol. 3, ed. I. M. Varentsov & G. Grasselly, E. Schweizerbart'sche Verlagsbuchhandlung, Stuttgart, pp. 31–90.

LANCELOT, Y., HATHAWAY, J. C. & HOLLISTER, C. D. (1972). Lithology of sediments from the western North Atlantic, Leg 11, Deep Sea Drilling Project. *Initial Reports of the Deep Sea Drilling Project,* **11,** 901–73.

LARSEN, E. R. & BERMAN, H. (1943). The microscopic determination of the non-opaque minerals. *Bulletin of the US Geological Survey*, **843**, 30–2.

LISITZIN, A. P. (1972). Sedimentation in the World Ocean. *Society of Economic Paleontologists and Mineralogists Special Publication*, **17**, 218 pp.

LOGVINENKO, N. V. (1982). Origin of glauconite in the Recent bottom sediments of the Ocean. *Sedimentary Geology*, **31**, 43–8.

MacDOUGALL, D. (1971). Fission track dating of volcanic glass shards in marine sediments. *Earth and Planetary Science Letters*, **10**, 403–6.

McRAE, S. G. (1972). Glauconite. *Earth Science Reviews*, **8**, 397–440.

MARSHALL, B. R. (1961). Devitrification of natural glass. *Geological Society of America Bulletin*, **72**, 1493–520.

MATHEWS, W. H. (1951). A useful method of determining approximate composition of fine-grained igneous rocks. *American Mineralogist*, **36**, 92–101.

MAZZULLO, J., MEYER, A. & KIDD, R. B. (1987). A new sediment classification scheme for the Ocean Drilling Program. *ODP Technical Note*, **8**, 19 pp.

MÉLIÈRES, F. (1979). Mineralogy and geochemistry of selected Albian sediments from the Bay of Biscay, Deep Sea Drilling Project, Leg 48. *Initial Reports of the Deep Sea Drilling Project*, **48**, 855–75.

MELLIS, O. (1954). Volcanic ash horizons in deep-sea sediments from the eastern Mediterranean. *Deep Sea Research*, **2**, 89–92.

MILLIMAN, J. D. (1974). *Marine Carbonates—Recent Sedimentary Carbonates, Vol. 1.* Springer-Verlag, New York.

MILLIMAN, J. D. & MEADE, R. H. (1983). World-wide delivery of river sediment to the oceans. *Journal of Geology*, **91**, 1–21.

MILNER, H. B. (1962). *Sedimentary Petrography (4th Edition), Vol. II.* George Allen and Unwin, London.

MIYASHIRO, A., SHIDO, A. F. & EWING, M. (1969). Composition and origin of serpentinites from the Mid-Atlantic Ridge near 25° and 30° north latitude. *Contributions to Mineralogy and Petrology*, **23**, 117–127.

MOLINA-CRUZ, A. & PRICE, P. (1977). Distribution of opal and quartz on the ocean floor of the subtropical southeastern Pacific. *Geology*, **5**, 81–4.

MORGANSTEIN, M. M. (1967). Authigenic cementation of scoriaceous deep-sea sediments west of the Society Ridge, South Pacific. *Sedimentology*, **9**, 105–18.

MORSE, J. W. & BERNER, R. C. (1979). Chemistry of calcium carbonate in the deep oceans. In *Chemical Modelling in Aqueous Systems*, ed. E. A. Jenne. American Chemical Society, Washington DC (American Chemical Society Symposium Series, 93), pp. 499–535.

MORTON, A. C. (1985). Heavy minerals in provenance studies. In *Provenance of Arenites*, ed. G. G. Zuffa. D. Reidel, Dordrecht, pp. 249–77.

MOSSOP, S. C. (1964). Volcanic dust collected at an altitude of 20 km. *Nature*, **203**, 824–7.

NAYUDU, Y. R. (1964). Volcanic ash deposits in the Gulf of Alaska and problems of correlation of deep sea ash deposits. *Marine Geology,* **1,** 194–212.

NEIHEISEL, J. (1965). Source and distribution of sediments at Brunswick Harbor and vicinity, Georgia. *US Army Coastal Engineering Research Centre Technical Memorandum,* **12,** 21 pp.

NELSON, C. H. & HOPKINS, D. M. (1972). Sedimentary processes and distribution of particulate gold in the northern Bering Sea. *US Geological Survey Professional Paper,* **689,** 27 pp.

NESTEROFF, W. D. (1973). Mineralogy, petrography, distribution and origin of the Messinian Mediterranean evaporites. *Initial Reports of the Deep Sea Drilling Project,* **13,** 673–94.

NINKOVICH, D. (1979). Distribution, age and chemical composition of tephra layers in deep-sea sediments off Western Indonesia. *Journal of Volcanology and Geothermal Research,* **5,** 67–86.

NINKOVICH, D., HEEZEN, B. C., CONOLLY, J. R. & BURCKLE, L. H. (1964). South Sandwich tephra in deep-sea sediments. *Deep Sea Research,* **11,** 605–19.

O'KEEFE, J. A. (1970). Tektite glass in Apollo 12 sample. *Science,* **168,** 1209–10.

ODEGARD, M. E. (1981). A new insight into the origin of dolomite on the ocean floor. *Naval Research Reviews,* **33,** 37.

ODIN, G. S. & STEPHAN, J. F. (1981). The occurrence of deep water glaucony from the Eastern Pacific: the result of in-situ genesis or subsidence. *Initial Reports of the Deep Sea Drilling Project,* **66,** 419–28.

OLAUSSON, E. (1961). Studies of deep-sea cores. *Reports Swedish Deep-Sea Expedition,* **8,** 337–91.

OVEREEM, Van, A. J. A. (1960). The geology of the cassiterite placers of Billiton, Indonesia. *Geologie En Mijnbouw,* **39,** 444–57.

PETERSON, M. N. A. & GOLDBERG, E. D. (1962). Feldspar distributions in South Pacific pelagic sediments. *Journal of Geophysical Research,* **67,** 3477–92.

PETTIJOHN, F. J. (1941). Persistance of heavy minerals and geologic age. *Journal of Geology,* **49,** 610–25.

PETTIJOHN, F. J. (1975). *Sedimentary Rocks (3rd Edition).* Harper and Row, New York.

PETZING, J. & CHESTER, R. (1979). Authigenic marine zeolites and their relationship to global volcanism. *Marine Geology,* **29,** 253–71.

PIMM, A. C., GARRISON, R. E. & BOYCE, R. E. (1971). Sedimentology synthesis: lithology, chemistry and physical properties of sediments in the northwestern Pacific Ocean. *Initial Reports of the Deep Sea Drilling Project,* **6,** 1131–252.

PITTMAN, E. D. (1963). Use of zoned plagioclase as an indicator of provenance. *Journal of Sedimentary Petrology,* **33,** 380–6.

PORRENGA, D. H. (1967). Glauconite and chamosite as depth indicators in the marine environment. *Marine Geology,* **5,** 495–501.
REVELLE, R. & EMERY, K. O. (1951). Barite concretions from the ocean floor. *Bulletin of the Geological Society of America,* **62,** 707–24.
REX, R. W. (1967). Authigenic silicates formed from basaltic glass by more than 60 million years contact with seawater, Sylvania Guyot, Marshall Islands. *Clay and Clay Minerals,* **15,** 195–203.
REX, R. W. & GOLDBERG, E. D. (1958). Quartz contents of pelagic sediments of the Pacific Ocean. *Tellus,* **10,** 153–9.
RINALDI, R., PLUTH, J. J. & SMITH, J. V. (1974). Refinement of crystal structure of phillipsite and harmotome. *Acta Crystallographica,* **30B,** 2426–33.
ROBERT, C. & CHAMLEY, H. (1974). Gypse et sapropels profounds de Mediterranée orientale. *Compte Rendu de l'Académie des Sciences,* **278D,** 843–6.
ROSS, D. A., NEPROCHNOV, Y. P., *et al.* (1978). *Initial Reports of the Deep Sea Drilling Project,* US Government Printing Office, Washington DC, **42** (2), 13–15.
ROTHE, P. & KOCH, R. (1978). Miocene volcanic glass from DSDP Sites 368, 369 and 370. *Initial Reports of the Deep Sea Drilling Project,* **41,** 1061–4.
ROZANOVA, T. V. (1971). The sediments of the rift zones in the mid-ocean ridges of the Indian Ocean. In *Istoriya Mirovogo Okeana* (*History of the World Ocean*), ed. L. A. Zenkevich. Nauka, Moscow, pp. 174–94 (in Russian).
RYAN, W. B. F., HSU, K. J. *et al.* (1973). *Initial Reports of the Deep Sea Drilling Project,* **13**. US Government Printing Office, Washington DC.
SAUNDERS, J. B., BERNOULLI, D., MULLER-MERZ, E., OBERHANSLI, H., PERCH-NIELSEN, K., RIEDEL, W. R., SANFILIPPO, A. & TORRINI, R. (1984). Stratigraphy of the late Middle Eocene to Early Oligocene in the Bath Cliff section, Barbados, West Indies. *Micropaleontology,* **30,** 390–425.
SCHALLREUTER, R. (1984). Framboidal pyrite in deep-sea sediments. *Initial Reports of the Deep Sea Drilling Project,* **75,** 875–91.
SCHEIDEGGER, K. F. (1973). Volcanic ash layers in deep sea sediments and their petrological significance. *Earth and Planetary Science Letters,* **17,** 397–407.
SCHEIDEGGER, K. F., JEZEK, P. A. & NINKOVICH, D. (1978). Chemical and optical studies of glass shards in Pleistocene and Pliocene ash layers from DSDP site 192, north-west Pacific Ocean. *Journal of Volcanology and Geothermal Research,* **4,** 99–116.
SCHMIDT, R. A. (1965). A survey of data on microscopic extraterrestrial particles. *NASA Technical Note, NASA TN,* **D-2719,** 132 pp.
SCHNITKER, D., MAYER, L. M. & NORTON, S. (1980). Loss of calcareous microfossils from sediments through gypsum formation. *Marine Geology,* **36,** M35–M44.

SCHOLZE, H. & MULFINGER, H. O. (1959). Der einbau des wassers in glasern. *Glastechnische Berichte,* **32,** 381–6.
SCHRÖTER, K., LÄUCHLI, A. & SIEVERS, A. (1975). Mikroanalytische identifikation von bariumsulfat-kristallen in den statolithen der rhizoide von *Chara fragilis,* Desv. *Planta,* **122,** 213–25.
SHAW, D. M., WATKINS, N. D. & HUANG, T. C. (1974). Atmospherically transported volcanic glass in deep sea sediments: theoretical considerations. *Journal of Geophysical Research,* **79,** 3087–94.
SHEPPARD, R. A. & GUDE, A. J. (1968). Distribution and genesis of authigenic silicate minerals in tuffs of Pleistocene Lake Tecopa, Inyo County, California. *US Geological Survey Professional Paper,* **597,** 38 pp.
SHEPPARD, R. A. & GUDE, A. J. (1969). Diagenesis of tuffs in the Barstow Formation, Mud Hills, San Bernardino County, California. *US Geological Survey Professional Paper,* **634,** 35 pp.
SHEPPARD, R. A., GUDE, A. J. & GRIFFIN, J. J. (1970). Chemical composition and physical properties of phillipsite from the Pacific and Indian Oceans. *American Mineralogist,* **55,** 2053–62.
SHVETSOV, M. S. (1954). Concerning some additional aids in studying sedimentary formations. *Byulleten Moskovskogo Obshchestva Ispytateley Prirody* (*Bulletin Moscow Society of Naturalists*), **29,** 61–6 (in Russian).
SIESSER, W. G. (1978). Petrology and geochemistry of pyrite and marcasite in DSDP Leg 40 sediments. *Initial Reports of the Deep Sea Drilling Project,* supplement to Volumes **38–41,** 767–75.
SIESSER, W. G. & ROGERS, J. (1976). Authigenic pyrite and gypsum in South West African continental slope sediments. *Sedimentology,* **23,** 567–77.
SINGER, A. (1984). The palaeoclimatic interpretation of clay minerals in sediments—a review. *Earth Science Reviews,* **21,** 251–93.
SLAUGHTER, M. & HAMIL, M. (1970). Model for deposition of volcanic ash and resulting bentonite. *Geological Society of America Bulletin,* **81,** 961–8.
STONECIPHER, S. A. (1976). Distribution of deep-sea phillipsite and clinoptilolite. *Chemical Geology,* **17,** 307–18.
STONECIPHER, S. A. (1977). Origin, distribution and diagenesis of deep-sea clinoptilolite and phillipsite. PhD thesis, Scripps Institution of Oceanography (unpublished).
STONECIPHER, S. A. (1978). Chemistry of deep-sea phillipsite, clinoptilolite and host sediments. In *Natural Zeolites,* eds L. B. Sand & F. A. Mumpton. Pergamon Press, Oxford, pp. 221–34.
SVERDRUP, H. U., JOHNSON, M. W. & FLEMING, R. H. (1942). *The Oceans—Their Physics, Chemistry and General Biology.* Prentice-Hall, New York.
TAKAHASHI, J. & YAGI, T. (1929). Peculiar mud-grains and their relation to the origin of glauconite. *Economic Geology,* **24,** 838–54.
TERRY, R. D. & CHILINGAR, G. V. (1955). Summary of 'Concerning some

additional aids in studying sedimentary formations' by M. S. Shvetsov. *Journal of Sedimentary Petrology,* **25,** 229–34.

THOMPSON, G., BOWEN, V. T., MELSON, W. G. & CIFELLI, R. (1968). Lithified carbonates from the deep-sea of the equatorial Atlantic. *Journal of Sedimentary Petrology,* **38,** 1305–12.

TRIPLEHORN, D. M. (1966). Morphology, internal structure and origin of glauconite pellets. *Sedimentology,* **6,** 247–66.

UREY, H. C. (1971). Tektites from the Earth. *Science,* **171,** 312–3.

VALLIER, T. L. & KIDD, R. B. (1977). Volcanogenic sediments in the Indian Ocean. In *Indian Ocean Geology and Biostratigraphy,* ed. J. R. Heirtzler *et al.* American Geophysical Union, Washington DC, pp. 87–118.

WENTWORTH, C. K. (1922). A scale of grade and class terms for clastic sediments. *Journal of Geology,* **30,** 377–92.

WEYL, W. A. (1951). Light absorption as a result of the interaction of two states of valency of the same element. *Journal of Physical and Colloid Chemistry,* **55,** 507–12.

WILLIAMS, H., TURNER, F. J. & GILBERT, C. M. (1954). *Petrography: An Introduction to the Study of Rocks in Thin Section.* W. H. Freeman, San Francisco.

WINDOM, H. L. (1969). Atmospheric dust records in permanent snowfields: implications to marine sedimentation. *Geological Society of America Bulletin,* **80,** 761–82.

WINDOM, H. L. (1976). Lithogenous material in marine sediments. In *Chemical Oceanography, Vol. 5,* ed. J. P. Riley & R. Chester. Academic Press, New York, pp. 103–35.

WINTERER, E. L., EWING, J. I., *et al.* (1973). Site 165. *Initial Reports of the Deep Sea Drilling Project,* **17,** 47–102.

WRIGHT, F. W. & HODGE, P. W. (1965). Studies of particles for extraterrestrial origin. 4. Microscopic spherules from recent volcanic eruptions. *Journal of Geophysical Research,* **70,** 3889–98.

XAVIER, A. & KLEMM, D. D. (1979). Authigenic gypsum in deep-sea manganese nodules. *Sedimentology,* **26,** 307–10.

APPENDICES

APPENDIX 1: GLOSSARY OF DESCRIPTIVE TERMS USED IN THE TEXT

Acicular. Needle-like.

Acidic. A descriptive term used for an igneous rock containing more than 60% silica. It is sometimes loosely used as a synonym for felsic, although the two terms are not strictly equivalent.

Aggregate polarisation. The pattern seen when a fine-grained crystal aggregate is viewed in cross-polarised light. The particle shows a mixture of different interference colours due to its variously orientated component grains.

Allochem. A collective term for the various discrete, organised carbonate elements, of intrabasinal origin, that can serve as coarser framework grains in mechanically deposited limestones. Examples of allochems are ooids, pellets, intraclasts (redeposited limestone) and fossil fragments.

Anhedral. Showing no crystal form.

Anisotropic. Describes a substance having different physical properties in different directions. However, the term is usually used to denote variation in optical properties such as refractive index or colour absorption.

Asbestiform. Fibrous.

Basic. A descriptive term for igneous rocks having relatively low silica content (usually arbitrarily defined as containing 45–55% silica). Such rocks are relatively rich in iron, magnesium and/or calcium. Pyroxene and olivine are the common ferromagnesian minerals although hornblende and biotite may occur in small amounts.

Baveno Law. A form of twinning sometimes seen in orthoclase in which the form {021} is both the twin plane and composition plane.

Bladed. A flattened, elongate crystal.

Botryoidal. Resembling a bunch of grapes in form.

Carlsbad Law. A common form of twinning seen in orthoclase where {100} is the twin plane but {010} is the composition plane.

Composite grain. A sedimentary grain consisting of two or more discrete particles.

Composition plane. The plane where the individuals of a twinned crystal appear to come into contact.

Conchoidal fracture. A form of fracture which has the appearance of a curved concentrically ribbed surface like a bivalve shell. Typical of quartz and obsidian.

Contact twins. Twins where the twin individuals come into contact in a plane. This plane of contact is called the composition plane.

Cruciform. Shaped in the form of a cross.

Crystallite. A general term for any minute body of unknown composition or crystal form which does not polarise light. Crystallites represent the initial stage of the crystallisation of a magma.

Dipyramid. A closed crystallographic form having 6, 8, 12, 16 or 24 faces. It can be considered as being formed from pyramids by reflection across a horizontal mirror plane.

Dome. An open crystallographic form consisting of two non-parallel faces symmetrical with respect to a mirror plane.

Euhedral. Showing a fully developed crystal form.

Felsic. An acronymic word derived from *fel*spar and *si*lic and applied to igneous rocks having abundant light-coloured minerals in their mode.

Form. In crystallography, a group of crystal faces, all of which have the same relation to the elements of symmetry and have the same chemical and physical properties as they are underlain by like atoms of the same geometrical arrangement. Crystallographic forms are designated by the use of braces { }. Otherwise, used as a synonym of shape, although strictly refers to those aspects of shape that are not expressed by sphericity or roundness.

Fracture. Breakage of a mineral along a direction or directions which are not cleavage or fissility directions. Often has a characteristic appearance for the particular mineral concerned.

Framboid. Spheroidal grains composed of randomly or non-randomly arranged pyrite microcrystallites. Frequently found within microfossil tests. A common morphological form of pyrite.

Geniculate. A crystal twin which bends abruptly, e.g. crystals of rutile twinned on the second order pyramid {101}.

Habit. The shape of a crystal as determined by the relative development of individual faces and forms. Sometimes used as a general term for the outward appearance of a mineral.

Intermediate. A descriptive term for an igneous rock that is transitional between basic or acidic (or between mafic and felsic). Intermediate rocks generally have a silica content between 54 and 65%.

Isotropic. Describes a substance which has the same physical properties in all directions. Crystals are termed isotropic if their physical properties do not vary according to crystallographic direction (*cf.* anisotropic).

Lenticular. Resembling a double-convex lens in shape.

Lustre. A diagnostic property of minerals relating to the character of the light reflected from the minerals surface.

Mafic. A descriptive term for igneous rocks composed mainly of one or more ferromagnesian dark-coloured minerals in its mode.

Mammillary. Smooth rounded masses resembling breasts or portions of spheres.

Manebach Law. A form of twinning sometimes seen in orthoclase with {001} as the twin plane and composition plane.

Microlite. A microscopic crystal that polarises light and shows some identifiable optical properties.

Octahedral. Pertaining to an octahedron—an isometric (cubic) crystal form, comprising of eight faces, each of which is an equilateral triangle.

Ooid. A spherical or sub-spherical particle that has grown by accretion around a nucleus. The commonest type is calcareous and forms through inorganic precipitation in shallow wave-agitated water. A synonymous term is oolith.

Pellets. Small, usually rounded ovoid particles of sediment typically between 0·25 and 5 mm in length. Thought to be faecal in origin and excreted by a wide range of marine invertebrates, particularly worms, molluscs and echinoderms.

Penetration twins. A twinned crystal in which the individuals appear to have grown through one another.

Perlitic. A texture characteristic of glassy igneous rocks showing small spheruloids and cracks formed by contraction during cooling.

Perthitic feldspar. A variety of alkali feldspar in which there is intergrowth of two feldspar minerals. Strictly used, it denotes a potassium-rich phase (usually microcline or orthoclase) from which smaller amounts of a sodium-rich phase (usually albite) has exsolved.

Phenocryst. Relatively large crystal in a finer groundmass.

Pinacoid. An open crystallographic form made up of two parallel faces.

Polycrystalline. A number of crystals comprising a single grain.

Polysynthetic twinning. Repeated twinning of three or more individuals on the same twin law and on parallel composition planes. A common example is plagioclase twinning on the albite law (twin plane and composition plane {010}).

Prismatic. A descriptive term for a crystal whose habit is dominated by prism faces. Such crystals show one dimension markedly longer than the other two.

Pseudomorph. A mineral occurring in the crystal form of another mineral species. Commonly forms through alteration, substitution, incrustation or paramorphism.

Pumiceous. Having a texture characterised by numerous small cavities giving a spongy or frothy appearance.

Pyramid. An open crystallographic form composed of 3, 4, 6, 8 or 12 nonparallel faces that meet at a point.

Pyritohedron. A dodecahedral crystal form having 12 pentagonal faces that are not regular. Named after pyrite which can have this crystal form.

Rhombohedral. A parallelepiped crystal whose six identical faces are rhombs.

Scoria. Irregular, usually very vesicular bomb-sized pyroclasts.

Shard. Vitric pyroclastic fragment. Shards generally consist of bubble wall fragments derived from explosively disrupted pumice.

Spherulitic. A texture characterised by numerous spherulites—more or less spherical bodies consisting of radial masses of acicular crystals, commonly feldspar.

Subhedral. A grain showing some crystal form, intermediate between anhedral and euhedral.

Swallowtail twins. A form of twinning seen in some monoclinic crystals with {100} as both the twin plane and composition plane. Often seen in gypsum.

Tabular. A descriptive term for a crystal that shows one dimension markedly smaller than the other two.

Tritiform. Shaped like a grain of wheat.

Twin plane. The plane about which reflection twins appear as mirror images. It is parallel to an actual or possible crystal face but not to a plane of symmetry in the untwinned crystal.

Vermicular. Having a worm-eaten appearance or occurring in a worm-like form intergrown with another mineral.

Vesicle. Cavity formed by entrapment of a gas body.

Vesicular. Characterised by abundant vesicles.

Vitreous. Glassy.

APPENDIX 2: COMPARATIVE PERCENTAGE CHARTS FOR ESTIMATING PROPORTIONS OF SEDIMENTARY COMPONENTS

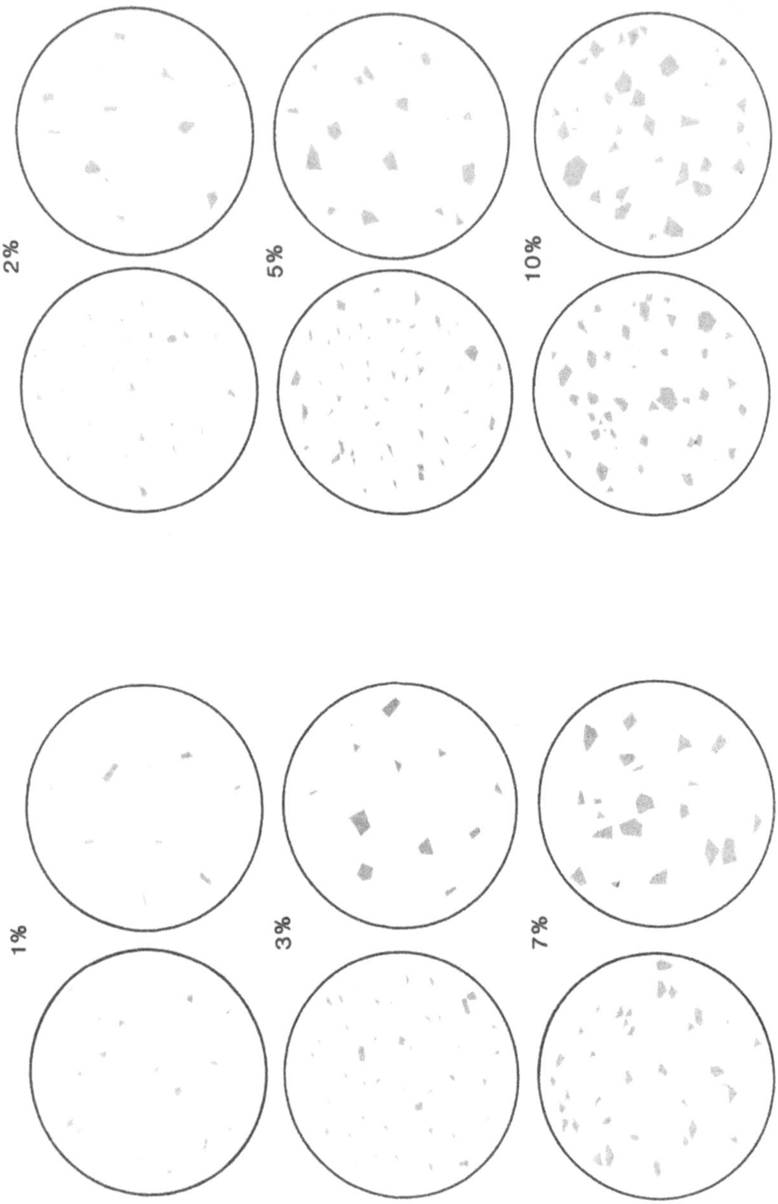

Fig. A2.1.

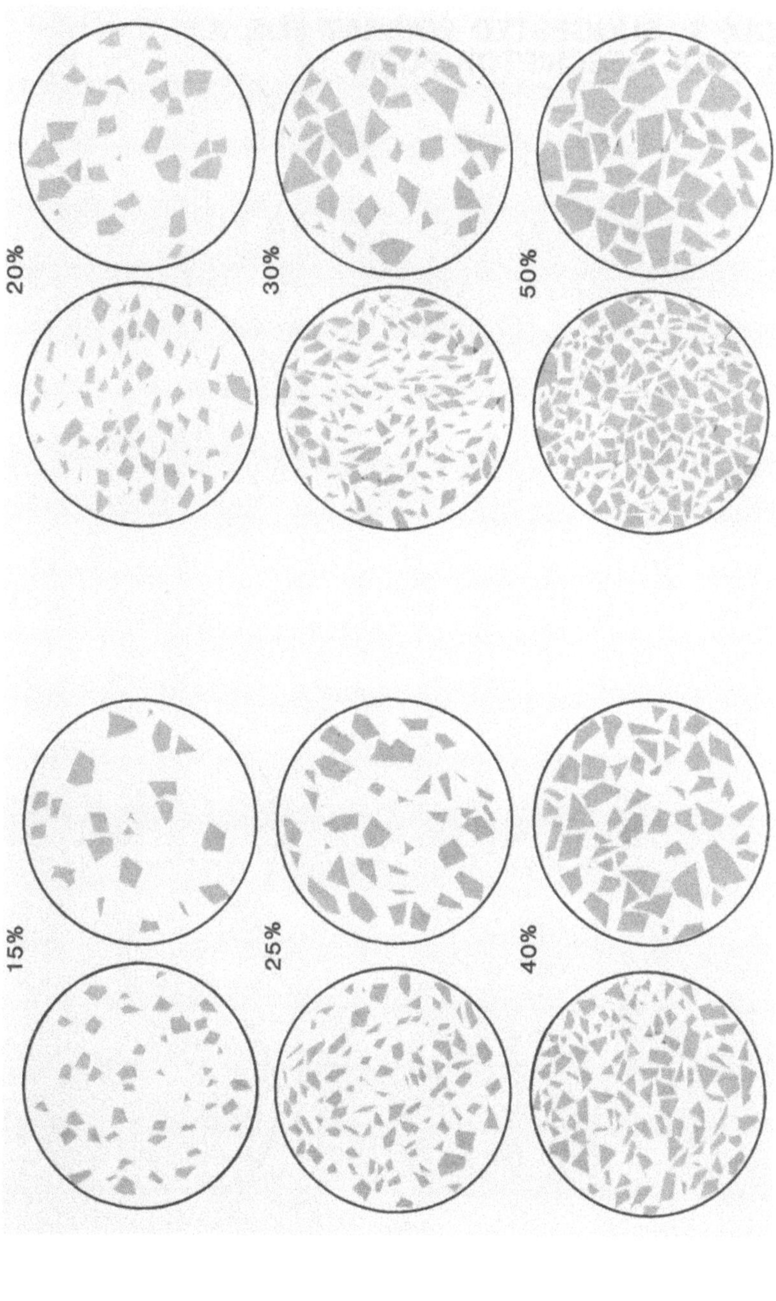

Fig. A2.2.

APPENDIX 3: SUGGESTED FORMAT FOR A SMEAR SLIDE DESCRIPTOR FORM

CORE SAMPLE DESCRIPTOR

ship.............. cruise................ leg............. station...........
lat............... long.................. water depth.........................
core n°........... section n°............ cm from top of core.................
cm from top of section..............
observer........................... date..................................

sediment type...
colour..$CaCO_3$..............%
remarks...
....
....

description of the residue:
abundance...
detrital minerals: authigenic materials:
quartz............ mica .... glauconite .... pyrite...........
feldspar.......... gypsum.......... Fe oxides............... arag.............
dark minerals...... calcite......... Micronodules............ dolomite.........
clay minerals...... Volcanic Glass.......... zeolite..........
other.............................. palag............

organic components:
bivalves........ gastropods......... shell fragments.......... bryozoa.........
fish teeth...... plant debris....... corals................... algae ....
ostracods....... echinoid spines.... sponge spicules.......... radiolarians....
diatoms......... coccoliths......... discoasters.............. spores..........
micrascidites of tunicates.......... organic matter........... fish remains....
benthic forams...................... planktonic forams........................
other............................... silicoflagallates........................
.... pteropods................................

list of the most characteristic taxa and their abundance:
....
....
....
....
....

AGE:..
ZONE:...
FACIES:...
....
remarks...
....
....

APPENDIX 4: FLOW CHARTS FOR THE IDENTIFICATION OF MINERAL GRAINS SEEN IN SMEAR SLIDES

Note. These charts provide a guide only and suggest likely identities of mineral grains. Identification should always be confirmed by consulting the main text.

Viewing in plane-polarised light:

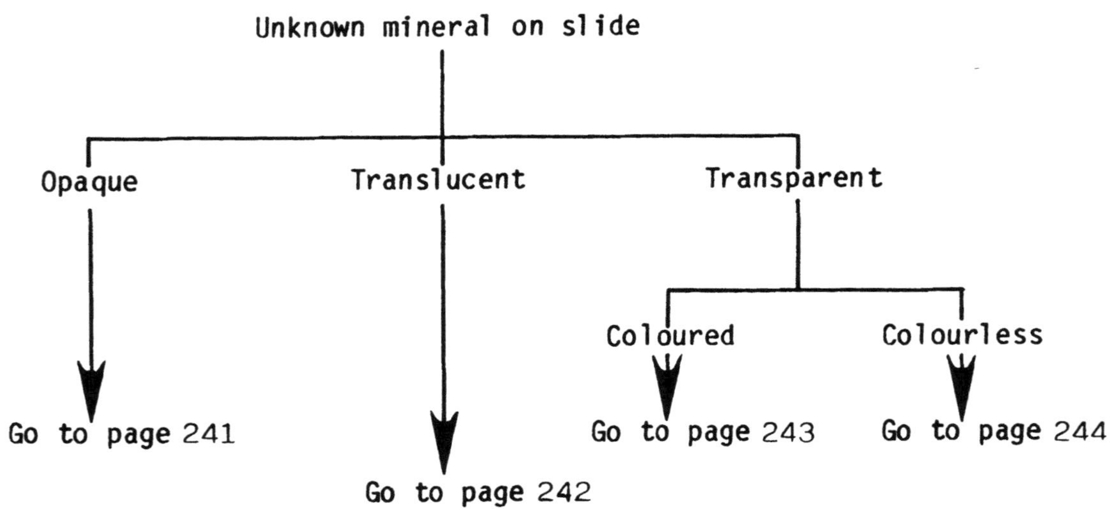

Viewing in plane polarised light:

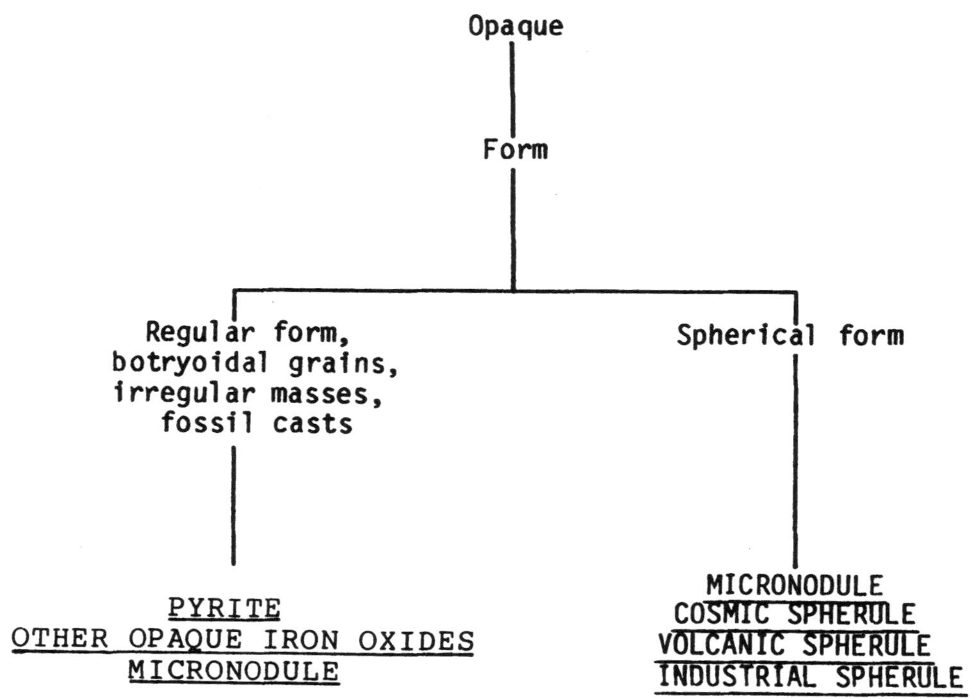

Viewing in plane polarised light:

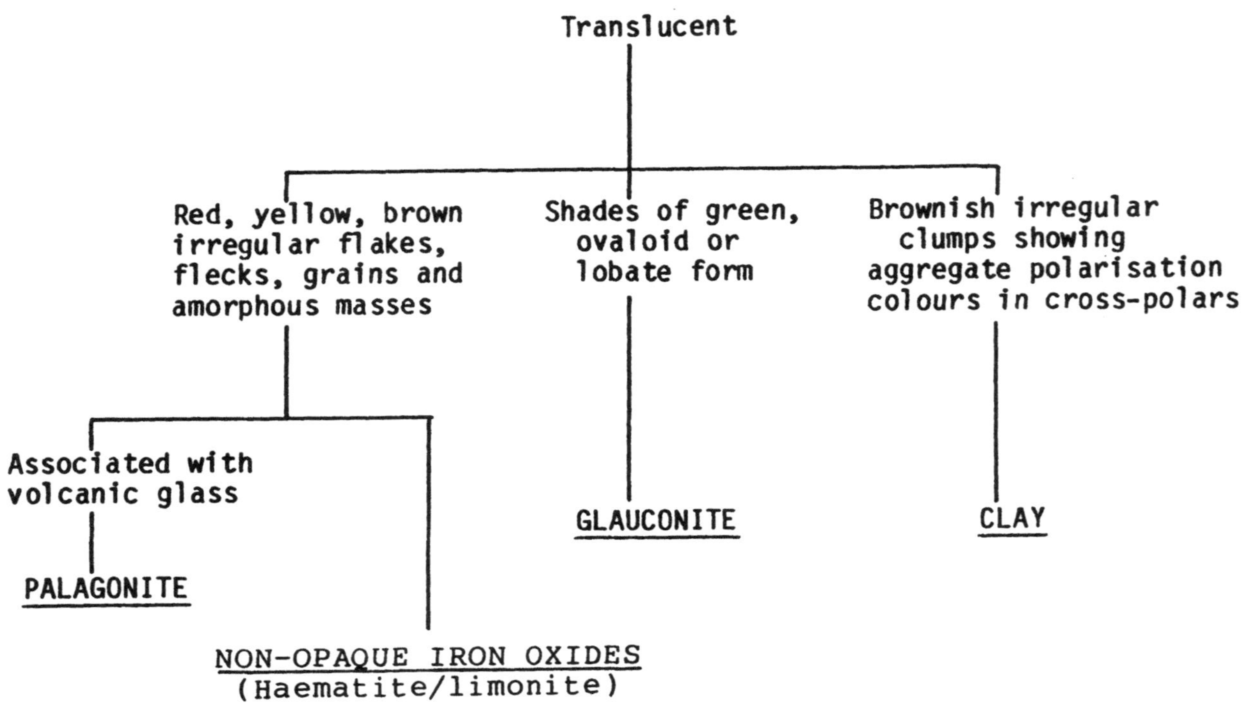

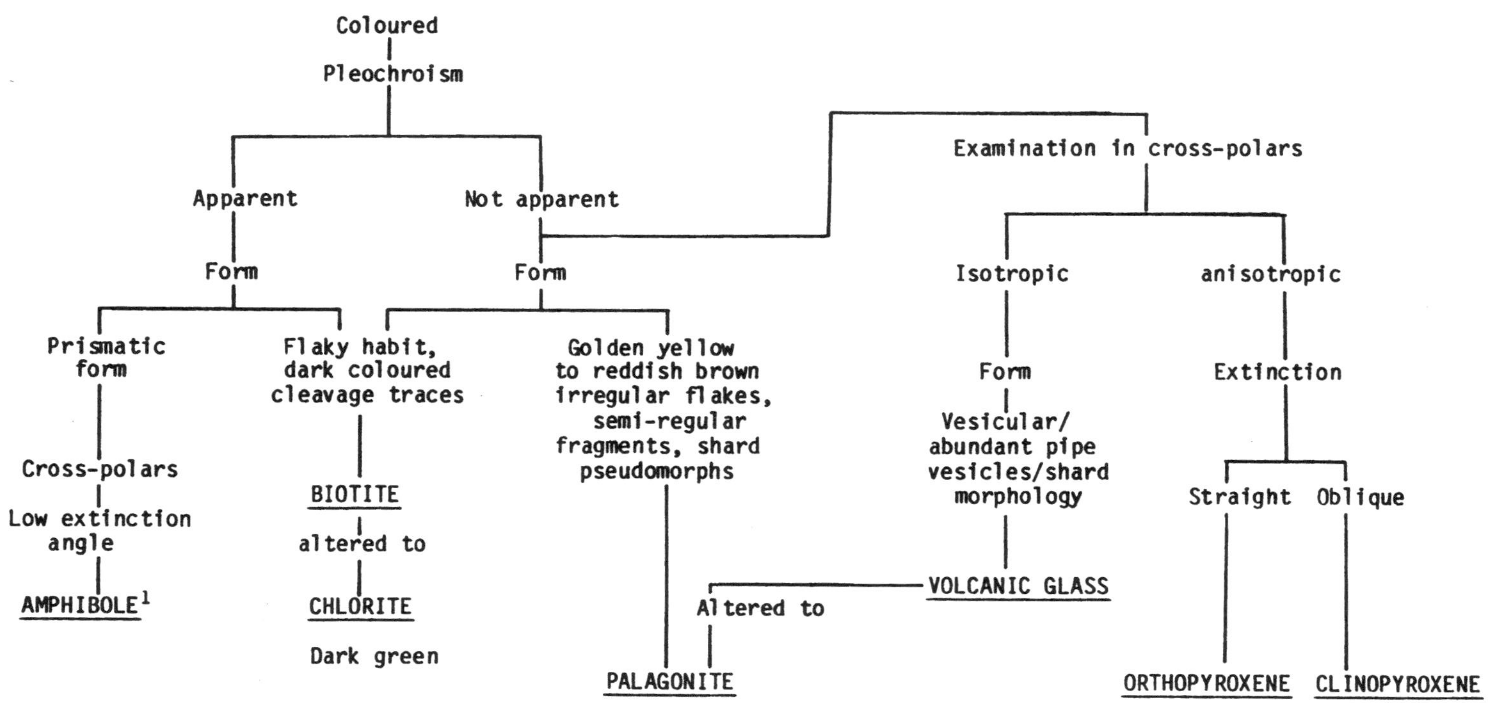

Note: [1] The clinopyroxene aegirine also has a small extinction angle and is coloured and pleochroic.

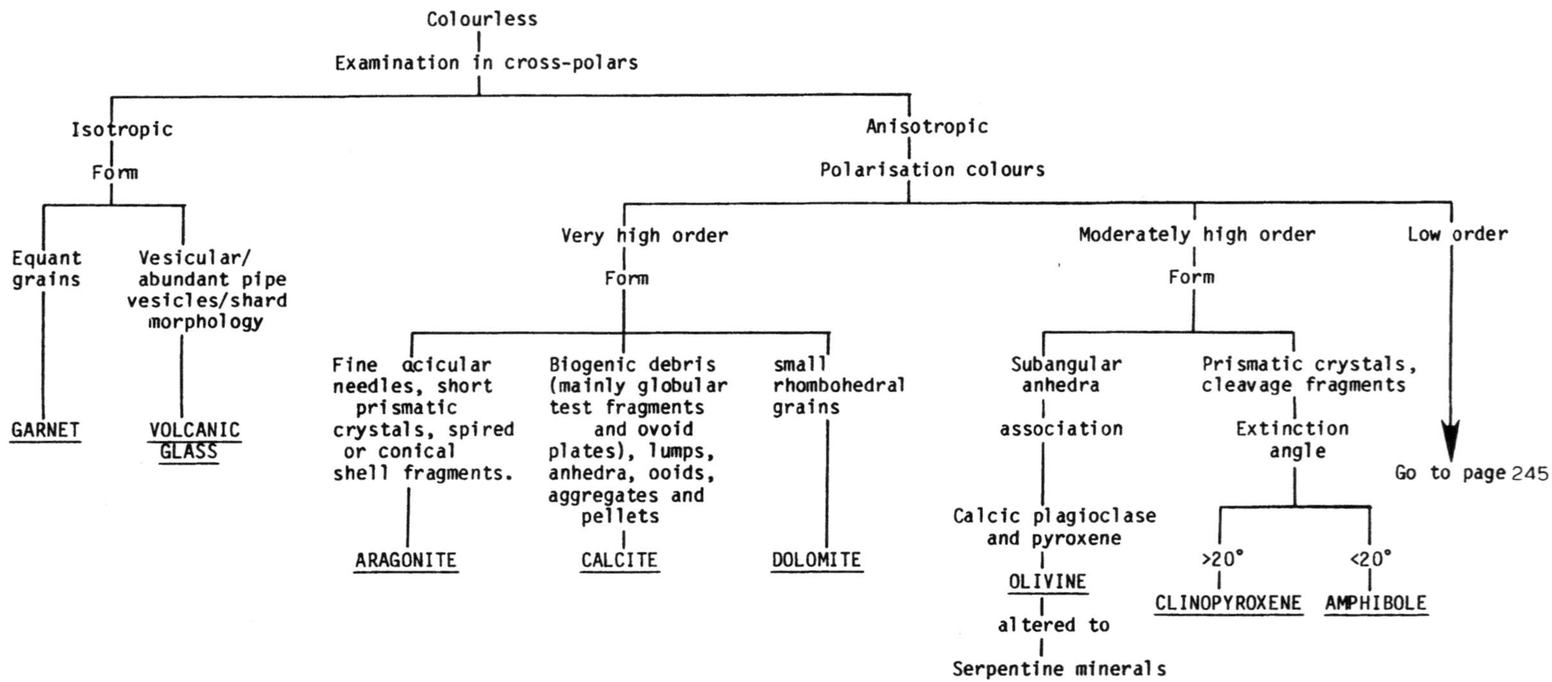
Colourless
Examination in cross-polars
Isotropic
Form
Equant grains
GARNET
Vesicular/ abundant pipe vesicles/shard morphology
VOLCANIC GLASS
Anisotropic
Polarisation colours
Very high order
Form
Fine acicular needles, short prismatic crystals, spired or conical shell fragments.
ARAGONITE
Biogenic debris (mainly globular test fragments and ovoid plates), lumps, anhedra, ooids, aggregates and pellets
CALCITE
small rhombohedral grains
DOLOMITE
Moderately high order
Form
Subangular anhedra
association
Calcic plagioclase and pyroxene
OLIVINE
altered to
Serpentine minerals
Prismatic crystals, cleavage fragments
Extinction angle
>20°
CLINOPYROXENE
<20°
AMPHIBOLE
Low order
Go to page 245

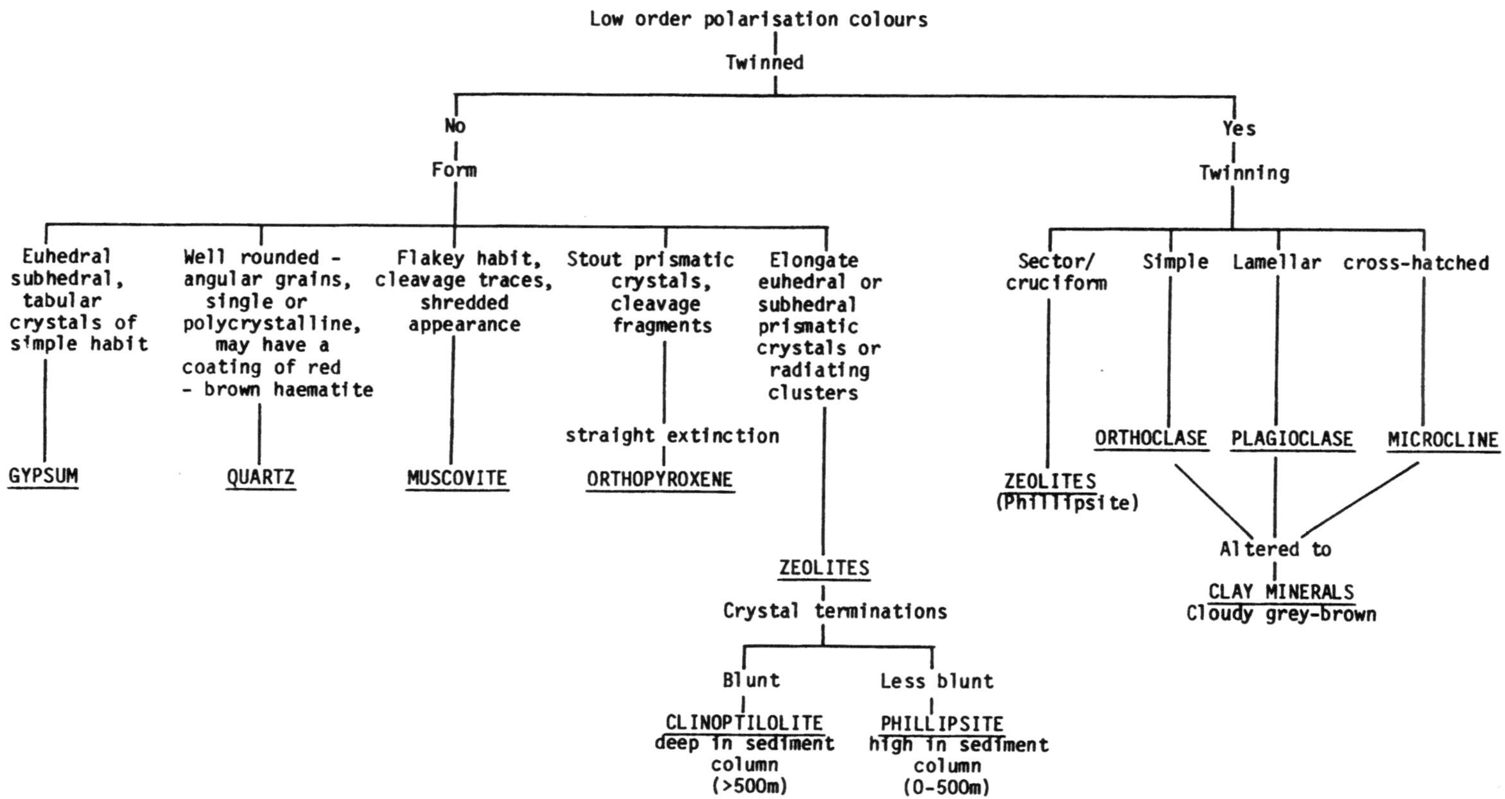
Low order polarisation colours
Twinned
No
Form
Euhedral subhedral, tabular crystals of simple habit
GYPSUM
Well rounded - angular grains, single or polycrystalline, may have a coating of red - brown haematite
QUARTZ
Flakey habit, cleavage traces, shredded appearance
MUSCOVITE
Stout prismatic crystals, cleavage fragments
straight extinction
ORTHOPYROXENE
Elongate euhedral or subhedral prismatic crystals or radiating clusters
ZEOLITES
Crystal terminations
Blunt
CLINOPTILOLITE
deep in sediment column (>500m)
Less blunt
PHILLIPSITE
high in sediment column (0-500m)
Yes
Twinning
Sector/ cruciform
ZEOLITES
(Phillipsite)
Simple
ORTHOCLASE
Lamellar
PLAGIOCLASE
cross-hatched
MICROCLINE
Altered to
CLAY MINERALS
Cloudy grey-brown

APPENDIX 5: AN ILLUSTRATED KEY TO THE IDENTIFICATION OF THE MAIN MICROFOSSIL GROUPS

Certain microfossil groups having hard parts capable of fossilisation are major contributors to marine sediments. Biogenic oozes are composed largely of the microscopic fossil tests of marine organisms and these may also occur in significant quantities in other types of marine sediments. These microfossil groups are of major importance in that much of what we know of oceanic palaeoenvironmental history has been derived from their study.

Although this book has been concerned with the recognition of mineral grains likely to be seen in marine sediments, it is important that sedimentologists should be able to recognise the major microfossil groups likely to be encountered in the analysis of smear slides. Hence, the inclusion of this appendix. Micropalaeontology is, of course, a specialist study and there are a number of excellent standard texts (for example, Glaessner, 1972; Haq & Boersma, 1978; Brasier, 1980; Bolli *et al.*, 1985). Therefore in this appendix, fossils are not identified to the generic or specific level but merely to their major groups, e.g. foraminifera, nannofossils, radiolaria, diatoms, sponge spicules, pollen and plant debris, etc. This will enable users to be at least aware of what these groups look like and enable them to classify a sediment through smear slide examination.

The following groups are briefly discussed and illustrated:

(a) Calcareous microfossils
- (1) Foraminifera
- (2) Coccolithophorids
- (3) Discoasters
- (4) Pteropods
- (5) Ostracods

(b) Siliceous microfossils
- (1) Radiolaria
- (2) Diatoms
- (3) Silicoflagellates
- (4) Sponge spicules

(c) Microscopic remains of larger organisms that may be seen in smear slides

(1) Phosphatic and fish debris
(2) Mollusc and echinoid fragments
(3) Plant debris, pollen and spores

Foraminifera

The Foraminifera are a group of protozoans belonging to the Class Sarcodina, Order Foraminiferida. They are characterised by a test of one to many chambers composed of secreted calcite or agglutinated grains. In life, they are either planktonic, living in the top 200–300 m of the water column, or benthonic (bottom-dwelling). On death, their remains make a major contribution to marine sediments deposited above the calcite compensation depth, especially deep-sea calcareous oozes (Fig. A5.1–A5.3). Test sizes are generally in the size range 50 to 1000 μm.

Foraminifera tests can show a great variety of form and chamber arrangement. Test shape may range from simple unilocular tests

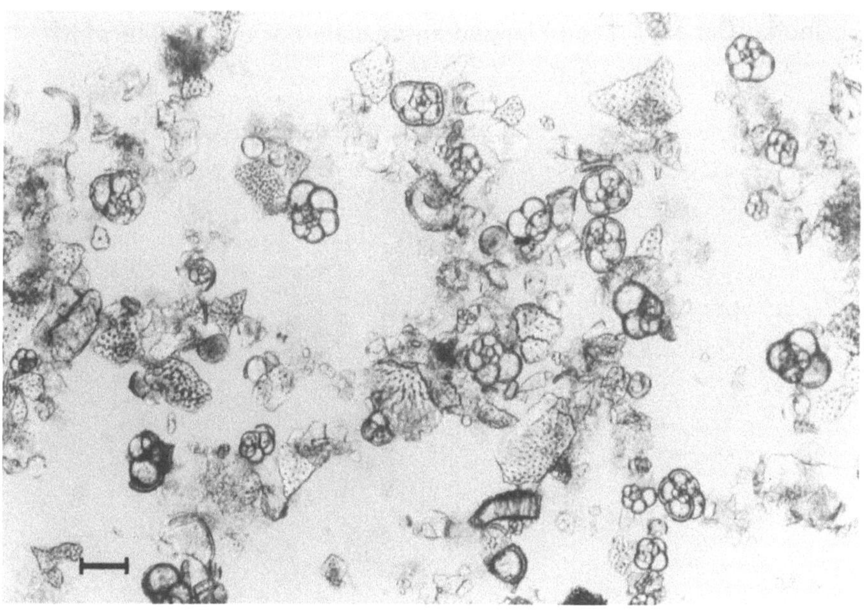

Fig. A5.1. Trochospiral foraminifers and larger perforate foraminifer test fragments. Pleistocene, Kings Trough, Northeast Atlantic. Plane-polarised light, scale bar equals 0·05 mm.

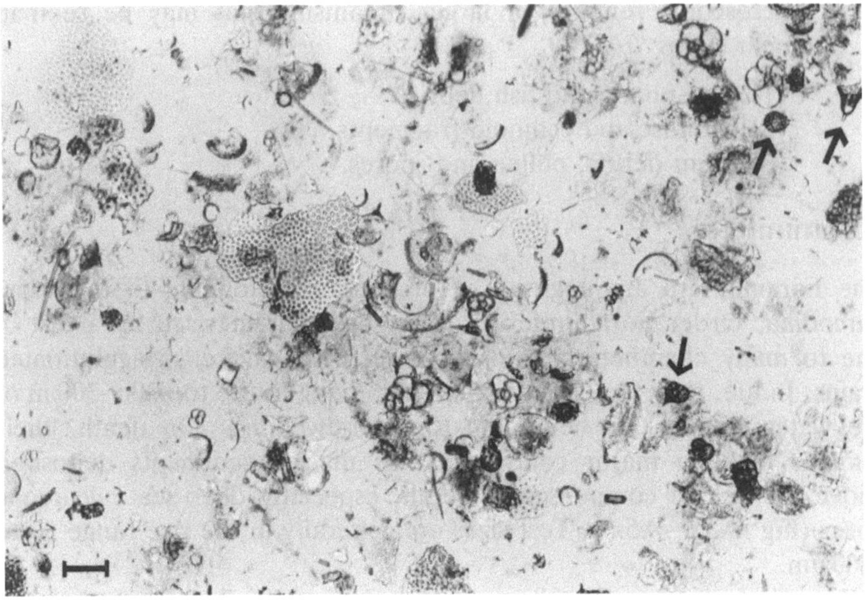

Fig. A5.2. Calcareous ooze comprising largely of foraminifera tests and test fragments. Some complete and fragmentary radiolarian tests also appear in the field of view (arrows). Pleistocene, Amirantes Passage, Indian Ocean. Plane-polarised light, scale bar equals 0·05 mm.

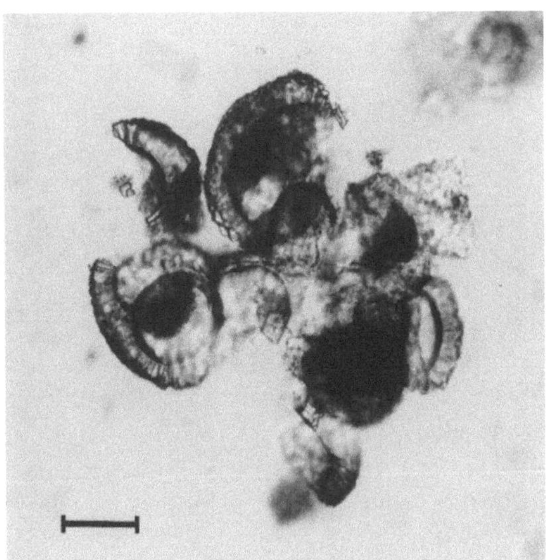

Fig. A5.3. Crushed foraminifer. Pleistocene, Kings Trough, Northeast Atlantic. Plane-polarised light, scale bar equals 0·05 mm.

through multilocular, planispiral (involute and evolute), trochospiral (high and low spired), and stretospiral (coiling in several planes). Chamber arrangement may be uniserial (i.e. arranged in a single row), biserial (arranged in two rows), triserial or even show a change in shape during growth (e.g. planispiral to uniserial). There may also be great variation in the size, shape, position and number of chamber apertures. There may be many different kinds of surface ornament. All these features are illustrated in standard texts.

Most of the tests that make a major contribution to calcareous deep-sea oozes are of planktonic foraminifera and all have calcareous tests. Tests of benthonic forms, which may be calcareous or composed of agglutinated grains, may be widespread but only form a minor constituent of deep-sea sediments. However, benthonic forms may be relatively common in terrigenous sediments on the continental shelf, slope and rise.

Foraminifera are of great geological importance due to their abundance, high diversity and occurrence in all marine environments. They are very important palaeoecological indicators and some forms have been used extensively as stratigraphic indices. They have been extensively used in palaeoceanographic analysis.

Range. Cambrian–Recent (planktonic forms first appeared in the late Jurassic).

Coccolithophorids

Coccoliths are minute, usually oval, calcite plates produced by unicellular planktonic algae belonging to the Phylum Chrysophyta, Class Haptophyceae. In life, the coccoliths, eight or more in number, depending on the species, are attached to a membrane surrounding a living cell. Each organism (i.e. the entire cell surrounded by coccolith plates) is termed a coccosphere. Coccospheres are usually spherical and generally 5–30 μm in diameter. The individual coccoliths are usually around 3 μm in diameter, although some forms have diameters as large as 35 μm. On the death of the organism, the membrane holding the coccolith plates breaks up and releases the individual coccoliths which then fall to the seafloor to make a major contribution to calcareous oozes. They form the dominant constituent of most nannofossil oozes and may, with clay, form a significant part of fine-grained terrigenous sediments.

The very small size of coccoliths means that high magnifications are

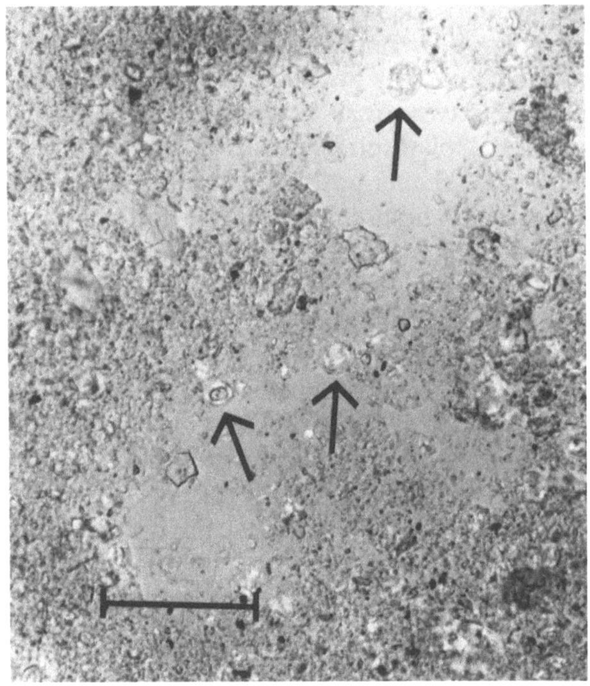

Fig. A5.4. Fine-grained calcite seen under high magnification. The material is largely composed of entire and broken coccolith plates. Three entire coccoliths are arrowed. Pleistocene, Kings Trough, Northeast Atlantic. Plane-polarised light, scale bar equals 0·05 mm.

often required to resolve individual plates (e.g. × 1000). Coccolith morphology forms the basis for the classification of the Coccolithophoridae, although in many species, a number of distinct types of coccolith can occur within the coccosphere. Determination of the fine details of coccolith morphology requires electron microscopy but most coccoliths can be recognised as distinct oval rings and plates using moderately high magnification with a light microscope (Figs A5.4–A5.6). They can certainly be observed using a ×10 or ×12·5 eyepiece and ×40 or ×100 objective, although the latter will require oil immersion. Under low magnification, coccolith oozes may resemble clay in plane-polarised light but will be easily be distinguished from clay by their high interference colours in cross-polars.

Coccoliths are valuable as stratigraphic indices and the Deep Sea Drilling Project relied heavily on this group of nannofossils for interpretations of sediment ages and the geological history of the ocean basins.

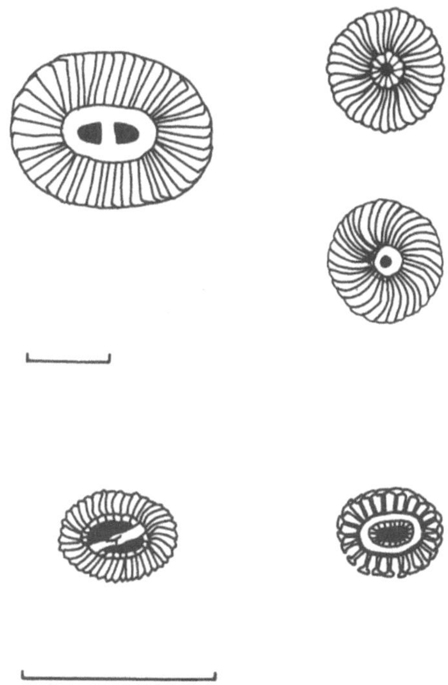

Fig. A5.5. Some Tertiary and Recent coccoliths. Scale bars equal 5 μm and refer to drawings directly above (after Black & Bukry in Fairbridge & Jablonski, 1979).

Range. Jurassic–Recent.

Discoasters

Discoasters are tiny star or rosette-shaped calcareous plates, 10–35 μm in diameter (Fig. A5.7). They are now extinct but are generally thought to be the remains of a coccolith-like planktonic organism. They are common in Tertiary deposits and appear to have become extinct around the Pliocene–Pleistocene boundary. Discoaster structure is relatively simple compared to coccoliths and the individual rays or arms are made of single tabular crystals. The great abundance, planktonic distribution pattern and rapid evolution of many species of *Discoaster* has made this group very important in oceanic Tertiary stratigraphy and in transoceanic stratigraphic correlation during this period.

Range. Late Palaeocene–late Pliocene.

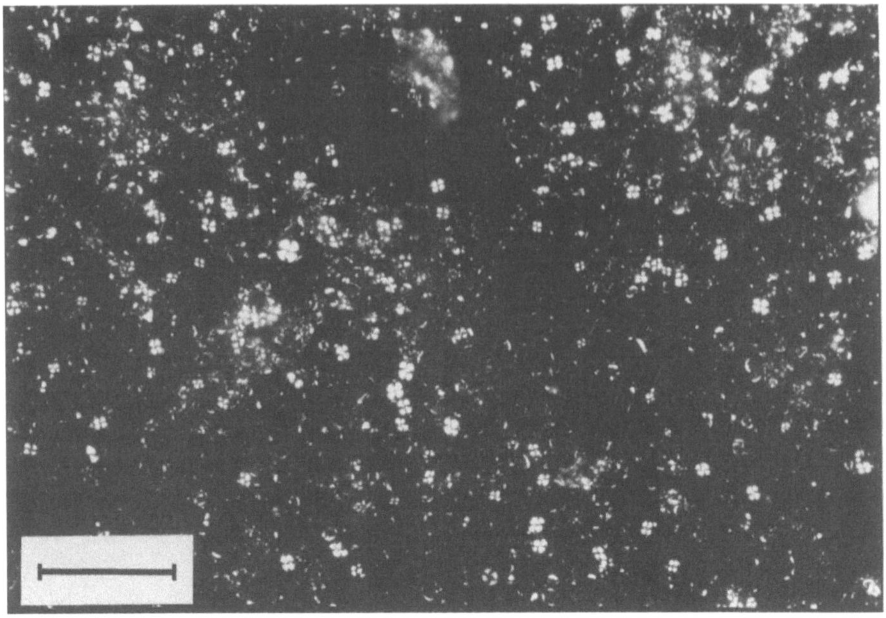

Fig. A5.6. Scatter of coccolith plates seen in cross-polarised light. Note the black interference crosses shown by each plate. Early Cretaceous, lower continental rise, southwest of New York, Northwest Atlantic. Scale bar equals 0·05 mm.

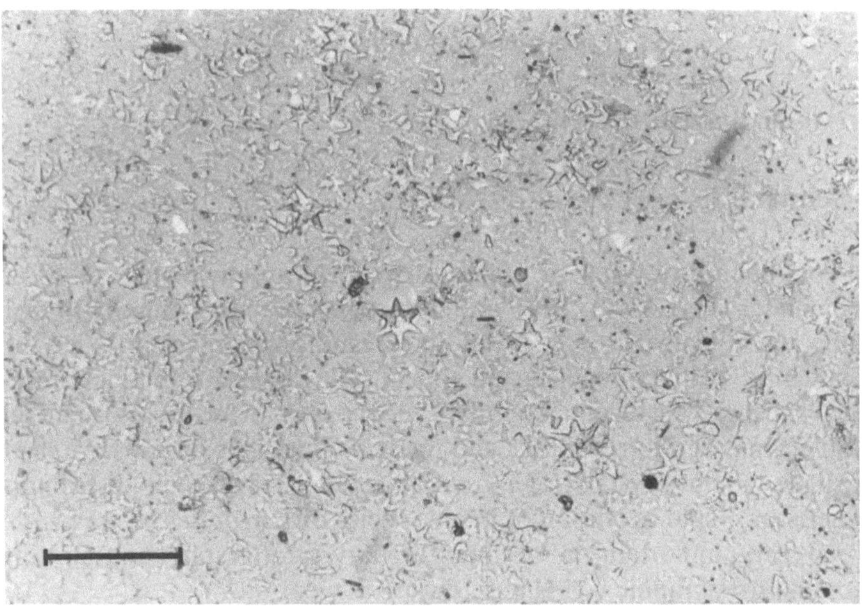

Fig. A5.7. Field of discoasters. Middle Eocene, Bay of Biscay, Northeast Atlantic. Plane-polarised light, scale bar equals 0·05 mm.

Pteropods

Pteropods are a group of planktonic opistobranch gastropods. They are relatively common zooplankton and some forms secrete delicate aragonitic shells (Figs. A5.8 and A5.9). The shells can show a great variety of shape, including elongate cones, spiral coils and trochospirals and delicate 'winged' forms. Shells may range up to 30 mm in length, although most are in the range 0·3 mm (300 μm) to 10 mm. Most pteropods live in the top few hundred metres of the water column, although there are a few bathypelagic species living at greater depths.

Although pteropods have a long Tertiary history, they are very rare in pre-Quaternary sediments due to the instability of aragonite, its susceptibility to dissolution and the fragility of the shells. They will only occur in sediments deposited above the aragonite compensation depth. Pteropod shells seen in smear slides will usually be broken and fragmentary.

Pteropods are of limited biostratigraphic value due to their rarity in pre-Quaternary sediments. However, as different species show distinct water mass preferences, particularly with regard to temperature, they have proved useful in Quaternary palaeoclimatic studies.

Range. ?Cretaceous–Recent.

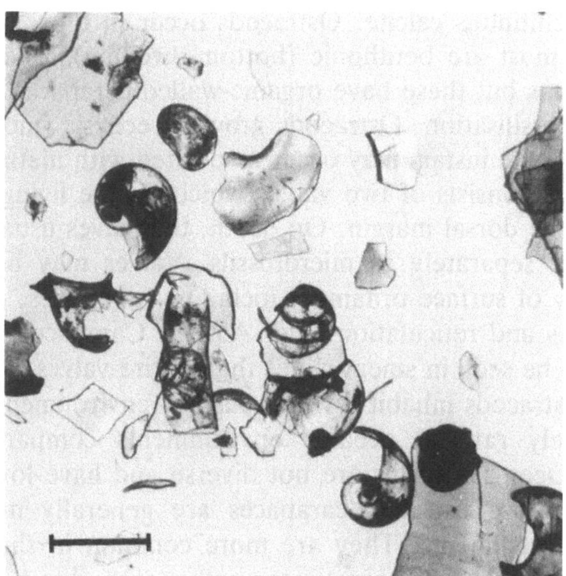

Fig. A5.8. Pteropod shells and shell fragments. Pleistocene, Red Sea. Plane-polarised light. Scale bar equals 0·5 mm.

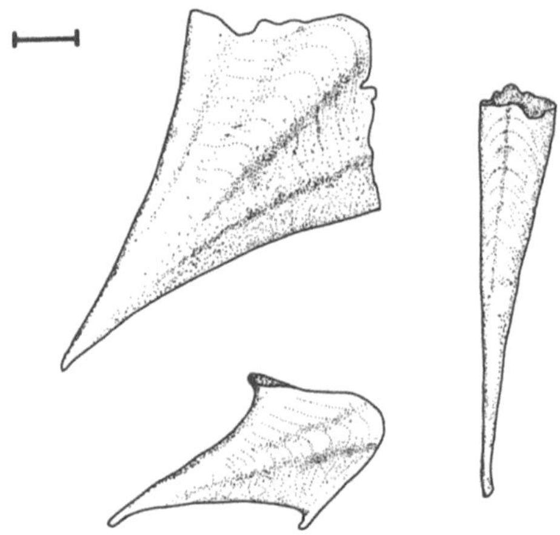

Fig. A5.9. Pteropod shells. Recent, Tuamotu Archipelago, Pacific Ocean. Scale bar equals 1 mm (after Reidel, 1963).

Ostracods

Ostracods are small bivalved crustaceans. Compared to most microfossils, they are relatively large, most being within the range 0·15 to 3 mm. The bivalved carapace is typically ovate, or bean or kidney-shaped and composed of chitinous calcite. Ostracods occur in brackish, fresh and seawater and most are benthonic (bottom-dwelling). There are some planktonic forms but these have organic-walled carapaces and have no potential for fossilisation. Ostracods grow by ecdysis (molting) so the remains of juvenile instars may occur associated with mature forms.

The carapace consists of two valves which, in the living animal, are hinged along the dorsal margin. On death, the valves usually disarticulate and occur separately as microfossils. Valves may be smooth or show a variety of surface ornament including tubercules, pores, lobes, ridges, grooves and reticulation (Fig. A5.10). Carapace fragments are more likely to be seen in smear slides than entire valves.

Although ostracods inhabit a wide range of environments, they tend to be relatively rare in oceanic environments compared to other microfossils. Deep-sea forms are not diverse and have low population densities, therefore ostracod carapaces are generally uncommon in most deep-sea sediments. They are more common in shallow marine sediments where they may sometimes number second to foraminifera in abundance among the microfossil fauna. Ostracods are useful for local or regional biostratigraphical zonation of shallow marine beds and can

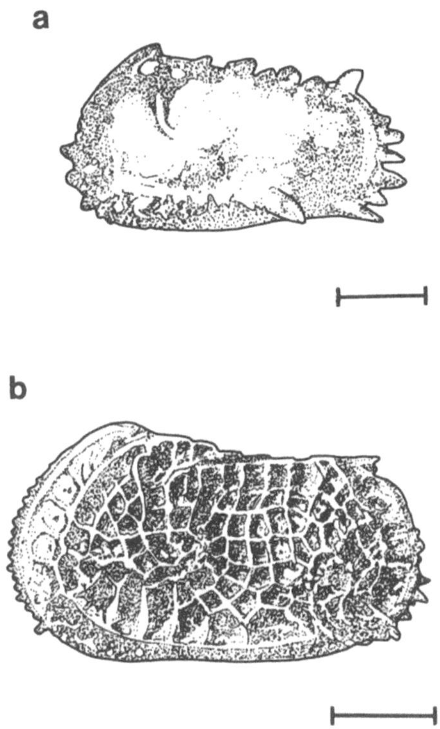

Fig. A5.10. Two deep sea ostracods: (a) *Pterygocythereis ceratoptera*; (b) *Bradleya dictyon*. Scale bars equal 0·1 mm and refer to drawing directly above (after Benson in Kennett, 1982).

be important indicators of temperature, salinities, seafloor depth and proximity to shorelines.

Range. Cambrian–Recent.

Radiolaria

Radiolaria are a diverse group of planktonic, pseudopod-bearing protozoans characterised by transparent opal skeletons. These are often exquisitely structured lattices and of great complexity (Figs A5.11 and A5.12). Radiolarians show a great variety of shape but most are generally based on spherical, conical or helmet-shaped ground forms. In size, most are within the range 20 to 400 μm.

Most radiolaria, like planktonic foraminifera, inhabit the upper 200–300 m of the water column, although some forms inhabit greater depths. However, they are rare in nearshore waters. Unlike diatoms,

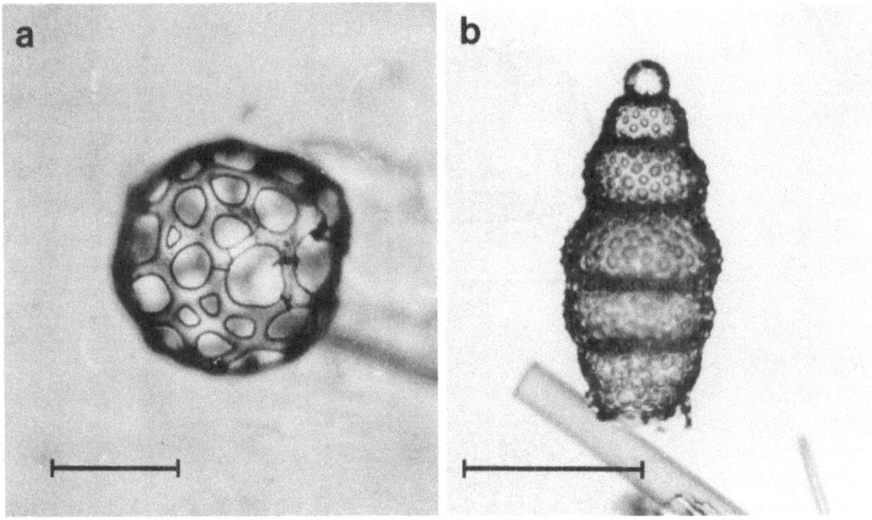

Fig. A5.11. Radiolarians: (a) *Collosphaera polygona* (Middle Miocene); (b) *Stichocorys peregrina* (Late Miocene). Both from DSDP Site 594, Southwest Pacific. Scale bars equal 0·05 mm. Both viewed in plane-polarised light. (Courtesy of Dr J.-P. Caulet and the Deep Sea Drilling Project.)

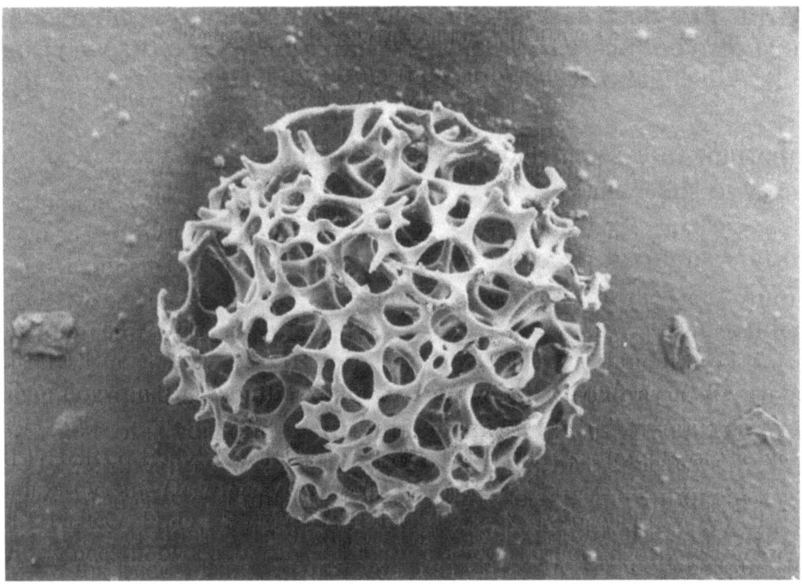

Fig. A5.12. Scanning electron micrograph of a spherical radiolarian. Radiolarian is 300 μm across. Quaternary, Indian Ocean. (Courtesy of Dr P. P. E. Weaver.)

there are no freshwater or benthonic forms. They are particularly abundant and diverse in equatorial latitudes, especially in areas of upwelling. However, they are also common in subpolar seas. Below the calcite compensation depth (typically 3000–5000 m), siliceous oozes will tend to accumulate on the ocean floor as nearly all calcite tests will have dissolved. However, with increasing depth there is progressive dissolution of silica tests as seawater is very undersaturated relative to silica. Therefore at great depths only the most resistant parts of radiolarian tests will survive, although good preservation may occur due to rapid sinking within faecal pellets.

Radiolarians are major contributors to deep-sea siliceous oozes. Radiolarian ooze is the main sediment type below the zones of high productivity in the equatorial Pacific at 3000–4000 m depth.

Radiolarians are perhaps, the most diverse group of oceanic microfossils. They are valuable in the biostratigraphic correlation of Mesozoic and Tertiary deep-sea sediments and have some potential as palaeoenvironmental indicators.

Range. ?Cambrian–Recent.

Diatoms

Diatoms are single-celled algae belonging to the Class Bacillariophyceae (Phylun Chrysophyta) that secrete a test (called the frustule) of opaline silica. The frustule is composed of two valves that fit over each other, much like a pill box, and are connected by a thin silica band (called the girdle band). Frustules are frequently perforate; these perforations, called punctae, are sometimes aligned in lines and these may be separated by imperforate ridges. Diatoms range from 2 μm to over 2000 μm in size, although most fall within the range 10 to 100 μm. They occur in seawater and freshwater. Being plants they are restricted to the photic zone (0–200 m water depth) during life.

Diatoms show a great variety in shape but two major divisions are recognised—centric diatoms (the Centrales) and pennate diatoms (the Pennales). Centric diatoms have spherical, circular, oblong, triangular or cylindrical tests with radial or concentric sculpture (Fig. A5.13). Pennate diatoms are spindle, rod or wedge-shaped and are bilaterally symmetrical about a median line. Most planktonic diatoms are centric forms while most benthonic diatoms are pennate. However, in Antarctic seas, planktonic pennate forms occur. The most common diatoms seen in smear slides of deep-sea siliceous oozes are discoidal or

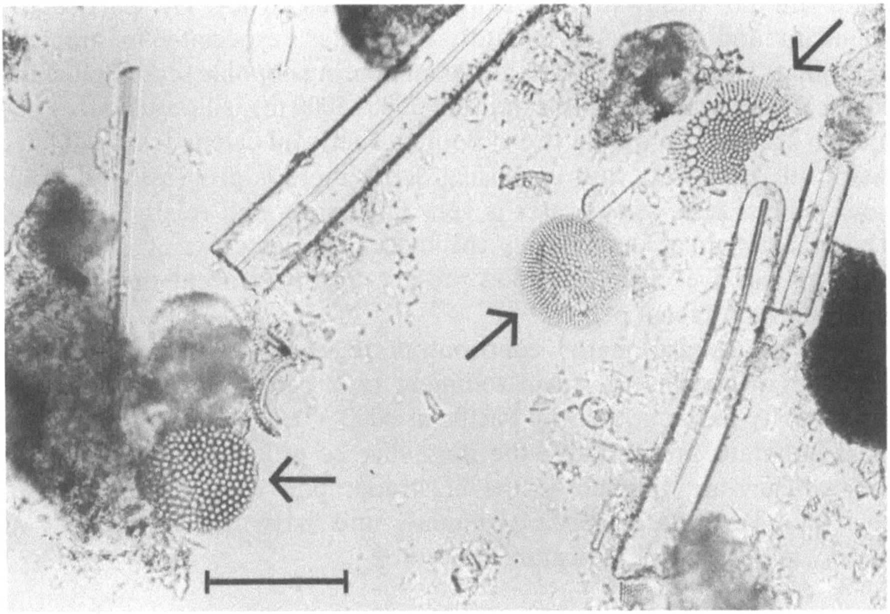

Fig. A5.13. Centric diatoms (arrowed). Some large siliceous sponge spicules are also present in the field of view. Note the central canals clearly visible within each spicule. Holocene, Angola Basin, Southeast Atlantic. Plane-polarised light, scale bar equals 0·05 mm.

elliptical centric forms; triangular or quadrangular forms are less common.

Diatoms represent most of the suspended silica in the water column and are the main contributors to deep-sea siliceous sediments. They are particularly abundant in regions of high productivity, particularly in high latitudes and areas of upwelling. Diatoms are important in the biostratigraphy of deep-sea sediments deposited below the calcite compensation depth. As different species have different requirements with regard to such factors as salinity and temperatures, diatoms are valuable in palaeoceanographic and palaeoecological studies. Like all planktonic microfossil groups, diatoms readily distinguish between near-surface cold and warm water oscillations. Hence they have been particularly valuable in Quaternary palaeoecological studies.

Range. ?Jurassic–Recent.

Silicoflagellates

Silicoflagellates are small, unicellular flagellated marine plankton having internal skeletons of opaline silica. Their true affinity is problematic, some workers regarding their means of movement as animal-like assign them to the Protozoa, whilst others regard them as plants since they contain photosynthetic chloroplasts. They possess an opaline skeleton of hollow rods arranged in a lattice, of which there are various types. These range from simple rings to relatively complex dome-like structures. A common arrangement is a basal ring from which rods arise on one side to form an arch or dome, resulting in an overall hemispherical shape (Fig. A5.14). Most silicoflagellates are in the range 20 to 50 μm and more rarely up to 100 μm.

Although silicoflagellates are widespread in sediments, they are seldom abundant and do not make a significant contribution to marine sediments. Usually no more than a few, if any, isolated examples will be noted in a smear slide. Although silicoflagellates are cosmopolitan, the distribution of different genera and species is closely controlled by water mass properties, especially temperature and salinity. Tempera-

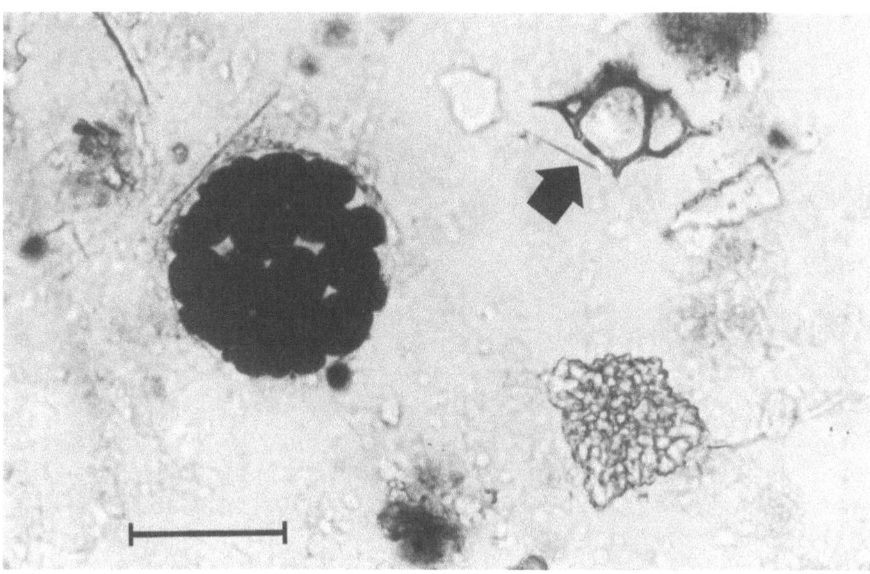

Fig. A5.14. Silicoflagellate test (arrowed). A centric diatom test containing framboidal pyrite is seen to the left. Holocene, Angola Basin, Southeast Atlantic. Plane-polarised light, scale bar equals 0·05 mm.

ture seems to be a particularly important control, resulting in a latitudinal differentiation in silicoflagellate assemblages. This has made them important in palaeoclimatic studies. Silicoflagellates are not particularly good zonal indices due to the slow evolutionary rate of the group. However, workable schemes have been devised and these may be useful for deep-sea sediments containing little or no carbonate.

Range. Cretaceous–Recent.

Sponge Spicules

Sponge spicules are minute siliceous or calcareous bodies occurring in, and serving to stiffen and support sponge tissue. Sponges are widespread on the seafloor and these contribute spicules to marine sediments on their death and breakup. Calcareous spicules are formed from crystalline calcite and may be needle-like or stellate with three–four radiating needle-like rays. Siliceous spicules are more common and may be needle-like, stellate with three or more rays, regular symmetrical or irregular in shape (Figs A5.15 and A5.16).

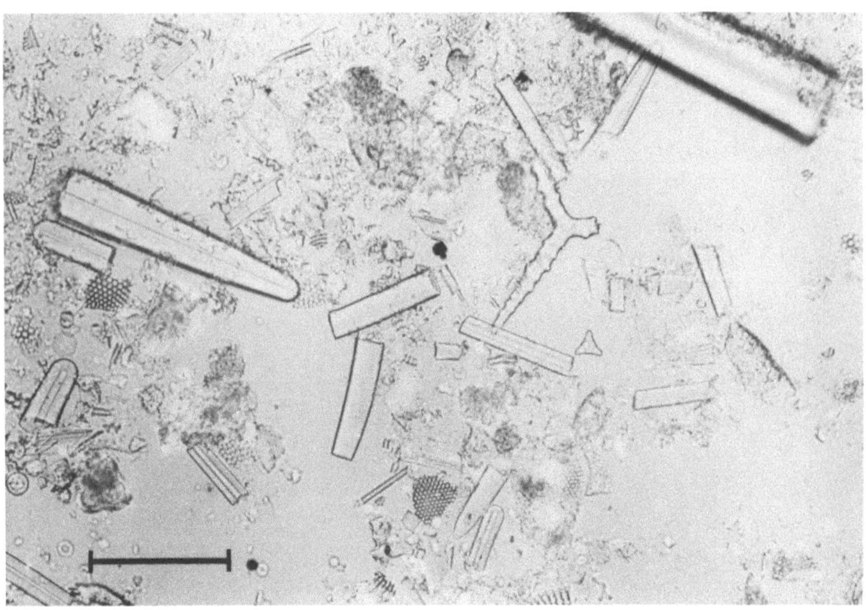

Fig. A5.15. Siliceous sponge spicules. Most are fragments of broken needle-like forms. Note the central canals. Holocene, Angola Basin, Southeast Atlantic. Plane-polarised light, scale bar equals 0·05 mm.

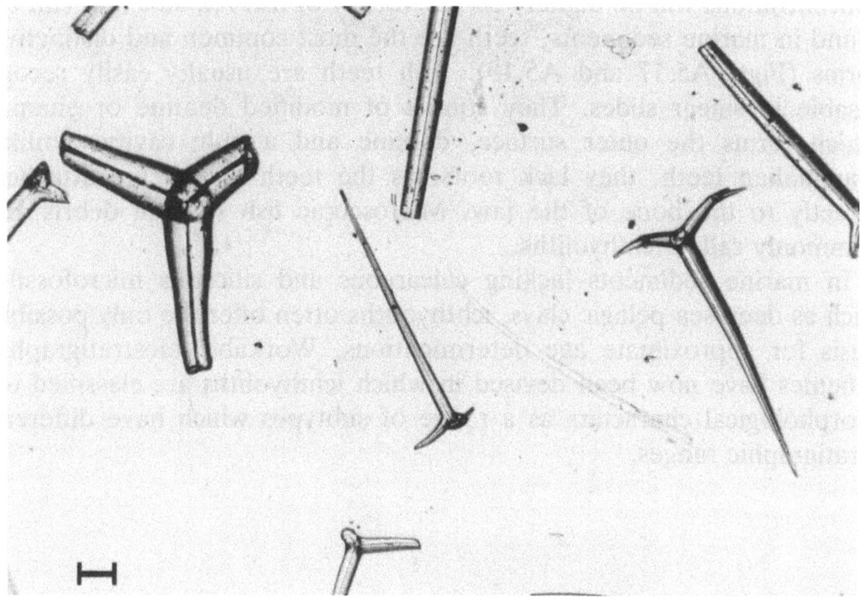

Fig. A5.16. Siliceous sponge spicules. Tyrrhenian Sea, Mediterranean. Plane-polarised light, scale bar equals 0·05 mm.

Names given to spicules are based on the number of rays or their symmetry. For example, needle-like spicules are termed monaxons while a spicule with four rays, tetrahedrally arranged, would be called a tetraxon. Size is very variable but sponge spicules may measure up to a few millimetres in length.

Siliceous sponge spicules can be distinguished from fragments of other siliceous microfossils by their generally smooth surface and by the presence of an internal axial canal. Although sponge spicules constitute only a minor proportion of deep-sea sediments, when they occur, they are frequently conspicuous in smear slides. They may occasionally form a significant part of some sediments on continental shelves or shallow banks.

Range. Cambrian–Recent.

Phosphatic and Fish Debris

Fragments of bone, small teeth and placoid scales form a minor constituent of all marine sediments. Although ubiquitous, they occur in low concentrations, except in areas of low sedimentation where concentrations will be higher. Of the variety of fish remains that can be found in marine sediments, teeth are the most common and distinctive forms (Figs. A5.17 and A5.18). Fish teeth are usually easily recognisable in smear slides. They consist of modified dentine or enamel which forms the outer surface, dentine and a pulp cavity. Unlike mammalian teeth, they lack roots, as the teeth of fish are attached directly to the bone of the jaw. Microscopic fish skeletal debris are commonly called ichthyoliths.

In marine sediments lacking calcareous and siliceous microfossils, such as deep-sea pelagic clays, ichthyoliths often offer the only possible basis for approximate age determinations. Workable biostratigraphic schemes have now been devised in which ichthyoliths are classified on morphological characters as a range of subtypes which have different stratigraphic ranges.

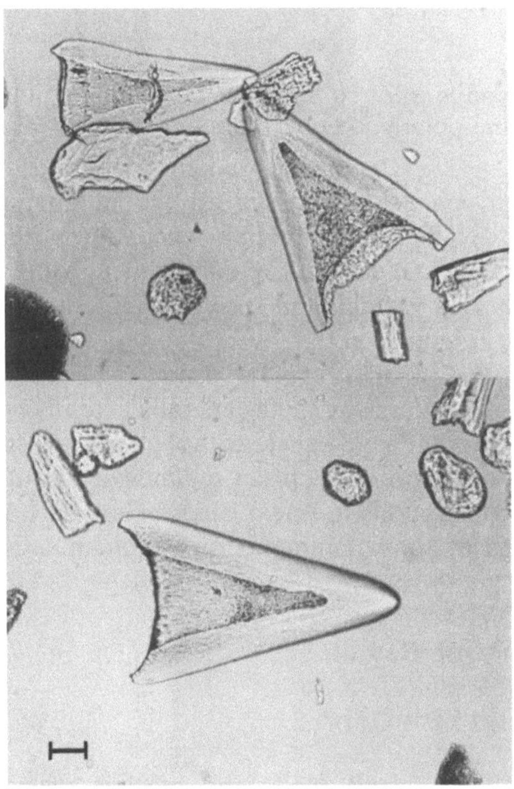

Fig. A5.17. Fish teeth. Paleogene, central North Pacific. Plane-polarised light, scale bar equals 0·05 mm. (Courtesy of Dr P. Doyle.)

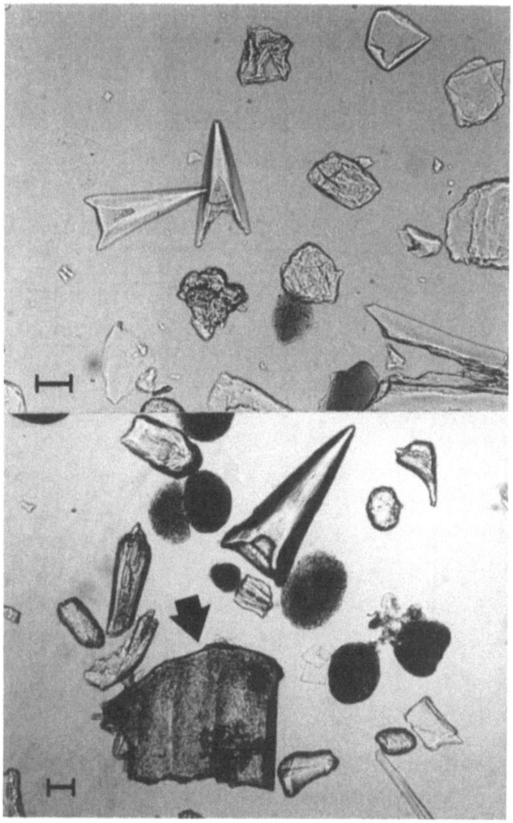

Fig. A5.18. Fish skeletal debris seen amongst mineral grains. In the bottom photograph a large piece of phosphatic skeletal material is seen (arrowed) together with a large tooth. Paleogene, central North Pacific. Plane-polarised light. Scale bar equals 0·05 mm. (Courtesy of Dr P. Doyle.)

Mollusc and Echinoid Fragments

Besides distinct microfossils, a wide variety of microscopic elements from larger animal skeletons may be seen in smear slides. Besides sponge spicules, already discussed, these may include alcyonarian spicules, echinoderm skeletal debris, molluscan shell fragments, early ontogenetic stages of echinoderms, molluscs and brachiopods, fragmentary bryozoa, small worm tubes and opercula and crustacean remains such as fragments of barnacle plates and carapaces. Generally, these have low abundances in most marine sediments but may be particularly common in some deposits.

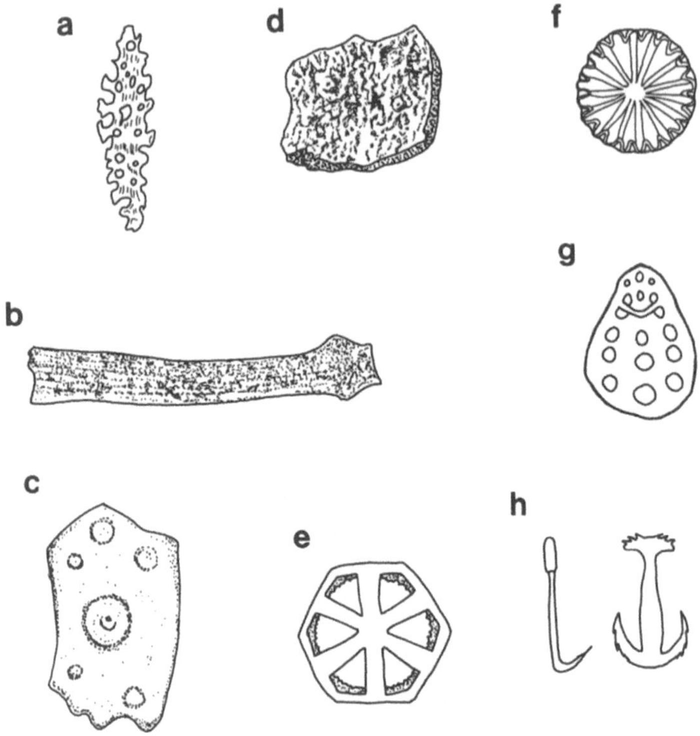

Fig. A5.19. Some microscopic elements from larger animal skeletons that may be seen in smear slides: (a) Alcyonarian spicule; (b) echinoid spine (original magnification × 10); (c) regular echinoid plate (original magnification × 6); (d) echinoid plate (original magnification × 13); (e)–(h) holothurian sclerites (original magnification × 150 approx.). ((a), (b), (d) and (f) after Glaessner, 1972; (e), (g) and (h) after Clarkson, 1979).

Alcyonarian corals possess microscopic calcareous or horny supportive elements called sclerodermites (Fig. A5.19a). These are commonly acicular and gently curved with granular surfaces. These may be locally relatively common in some marine sediments. Detached echinoderm skeletal elements that may be seen in smear slides include isolated terminal and marginal plates of asteroids (star fish), ophiuroid ossicles, plates and spines (brittle star remains) and crinoid fragments. Echinoids may be represented by isolated plates and spines (Fig. A5.19b–d). Plates may show tubercles, pores or other forms of ornament. Even small echinoid fragments can be recognised by their finely reticulate nature and distinctive crystallographic orientation of the calcite. Holothurians (sea cucumbers) are common in the deep sea and their body walls contain discrete calcareous spicules which are contributed to

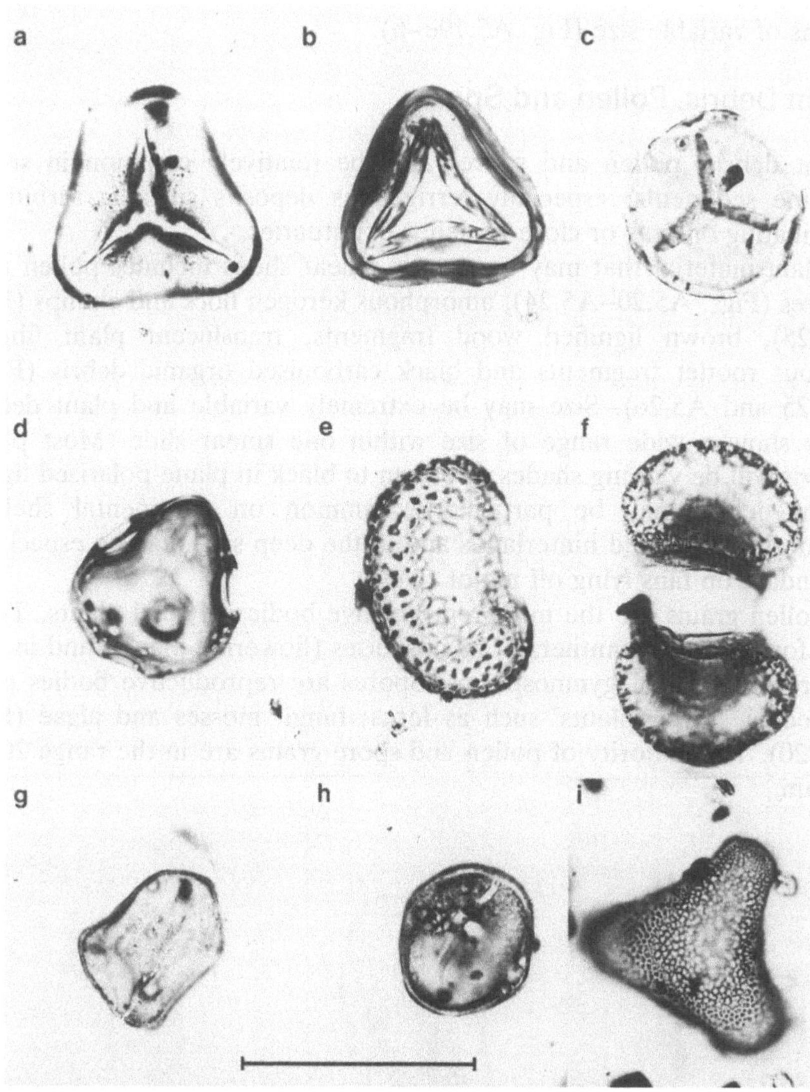

Fig. A5.20. Pollen and spores from DSDP Site 565, offshore western Costa Rica: (a)–(d) trilete fern spores; (e) monolete fern spore; (f) *Podocarpus* pollen;((g) Cyperacea pollen; (h) unknown tricolporate pollen grain; (i) Bombacaceae pollen. All Pleistocene. Plane-polarised light, scale bar equals 0·05 mm. (Courtesy of Dr Sally Horn, University of Tennessee and the Deep Sea Drilling Project.)

the sediment on death and decay of the organism. These may appear as minute reticulate plates and discs and wheel, anchor and hook-like forms of variable size (Fig. A5.19e–h).

Plant Debris, Pollen and Spores

Plant debris, pollen and spores may be relatively common in some marine sediments, especially terrigenous deposits such as turbidites originating on fans or close to deltas or estuaries.

Plant material that may be seen in smear slides includes pollen and spores (Figs. A5.20–A5.24), amorphous kerogen flocs and clumps (Fig. A5.25), brown lignified wood fragments, translucent plant fibres, fibrous rootlet fragments and black carbonised organic debris (Figs. A5.25 and A5.26). Size may be extremely variable and plant debris may show a wide range of size within one smear slide. Most plant debris will be varying shades of brown to black in plane-polarised light. Plant debris may be particularly common on continental shelves bordering vegetated hinterlands and in the deep sea, may be especially abundant on fans lying off major rivers.

Pollen grains are the male reproductive bodies of seed plants. They are formed on the anthers of angiosperms (flowering plants) and in the microsporangia of gymnosperms. Spores are reproductive bodies produced by 'lower plants' such as ferns, fungi, mosses and algae (Fig. A5.20). The majority of pollen and spore grains are in the range 20 to 80 μm.

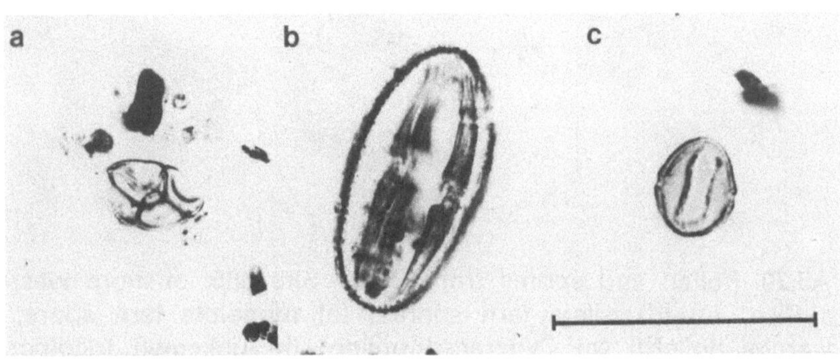

Fig. A5.21. Pollen grains from DSDP Site 565, offshore western Costa Rica: (a) Mimosaceae pollen tetrad; (b) *Spondias*-type (tricolporate) pollen; (c) Urticales pollen. Pleistocene, plane-polarised light, scale bar equals 0·05 mm. (Courtesy of Dr Sally Horn, University of Tennessee and the Deep Sea Drilling Project.)

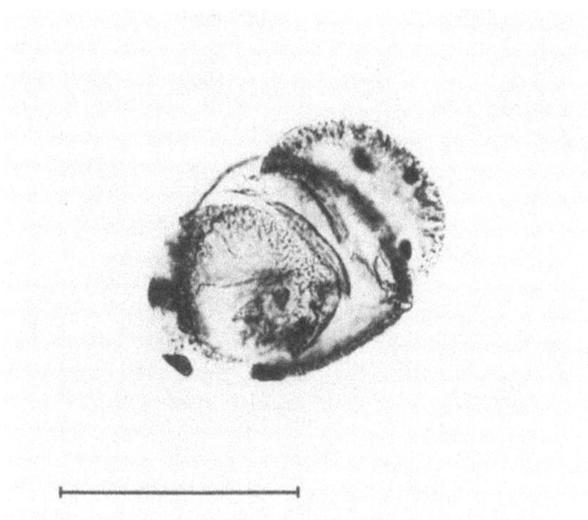

Fig. A5.22. *Pinus* pollen grain. Pleistocene, plane-polarised light, scale bar equals 0·05 mm. (Courtesy of Dr Sally Horn, University of Tennessee and the Deep Sea Drilling Project.)

Pollen may show a wide range of morphology. Grains may be subspheroidal, oblate, pentagonal, rod-shaped (prolate), vesiculate or saccate (possessing attached sacs, wings or bladders) or irregular in shape (Figs. A5.21–A5.23). They may show various types of opening on the grain surface. These are usually circular pores or elongate,

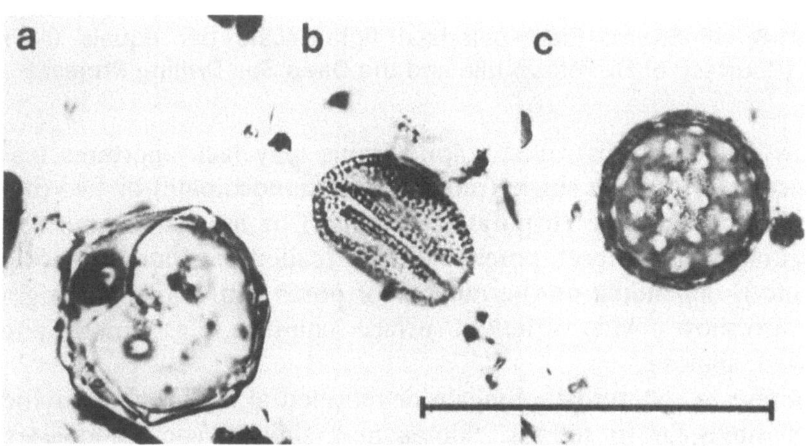

Fig. A5.23. Pollen from DSDP Site 565 offshore western Costa Rica: (a) *Juglans* pollen; (b) *Virola* pollen; (c) Amaranthaceae pollen. Pleistocene, plane-polarised light, scale bar equals 0·05 mm. (Courtesy of Dr Sally Horn, University of Tennessee and the Deep Sea Drilling Project.)

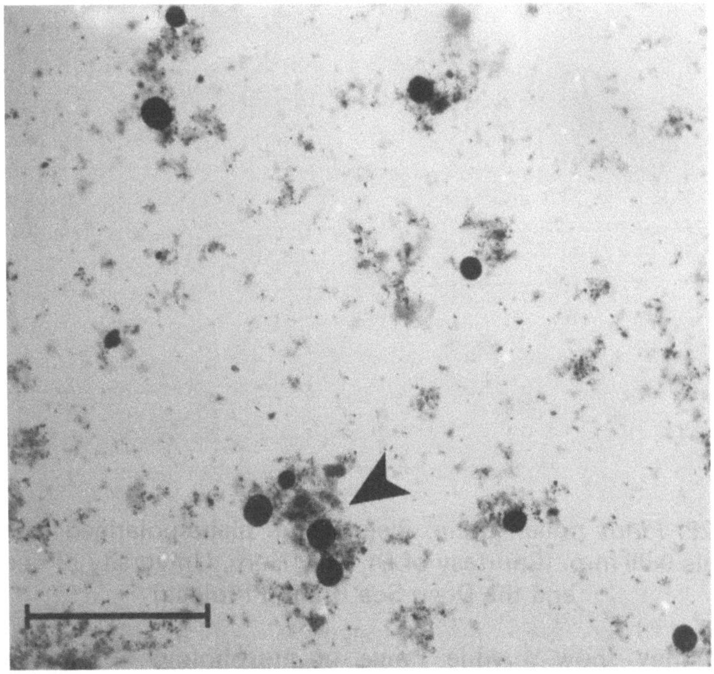

Fig. A5.24. Plant debris in Upper Miocene deep-sea carbonate mud from the Gardar Ridge, south of Iceland, showing abundant small flocs of amorphous debris and finely divided pyrite bodies and common pyritised (black) algal spores. A cluster of pyritised spores embedded in amorphous kerogen is seen at the bottom centre (arrow). DSDP Leg 94, 611-41-5, 48–50 cm. Plane-polarised light, scale bar equals 0·2 mm. (Courtesy of Dr Peta Mudie and the Deep Sea Drilling Project.)

furrow-like apertures called colpi. Grains may lack apertures (called inaperturate) or have single (monoporate/monocolpate) or two (diporate, dicolpate), three (triporate, tricolpate) or more apertures. Some grains may possess both pores and colpi (called dicolporate, tricolporate, etc.), depending on the number of pores and colpi. Pollen grains may also show a wide variety of surface sculpture, e.g. smooth, pitted, grooved, spinose.

Spores are spheroidal, elongate or tetrahedral bodies and may occur either singly or in tetrads. Spores may show single (monolete) or triradiate (trilete) surface sutures or may lack these (alete). Many spores show similar surface ornament as pollen grains. With regard to colour, both pores and pollen are usually various shades of brown or yellow in plane-polarised light.

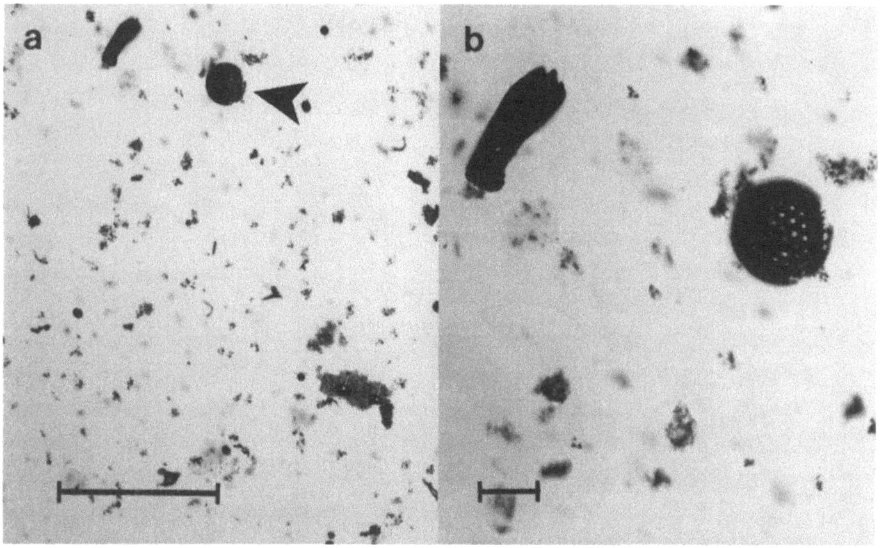

Fig. A5.25. (a) Plant debris in Upper Miocene deep-sea carbonate mud from the Gardar Ridge, south of Iceland, showing well dispersed small particles of amorphous kerogen, rare carbonised wood fragments and pyritised microfossil fragments (e.g. diatom frustule arrowed—see Fig. A5.25b). Scale bar equals 0·2 mm. (b) high magnification view of part of same sample showing pyritised diatom frustule, kerogen particles and carbonised wood fragment. Scale bar equals 0·02 mm. Both plane-polarised light, DSDP Leg 94, 611-44-1, 61–63 cm. (Courtesy of Dr Peta Mudie and the Deep Sea Drilling Project.)

Fluvial transport is probably the major means of pollen and spore dispersal to the marine environment, although wind transport can be locally important, particularly for small grains. However, pollen and spore counts in the atmosphere beyond the continental margins are low and therefore aeolian transport is probably of only minor importance in the transport of pollen to the deep sea.

Pollen and spores in marine sediments generally reflect coastal and regional vegetation. Pollen from vegetation zones deep in the continental hinterland may appear in marine sediments off major rivers. Quaternary and pre-Quaternary pollen and spores in marine sediments have been used as biostratigraphic tools and pollen profiles downcore have been used to elucidate changes in vegetation and associated environmental change onshore. Marine pollen may provide a direct means of correlating marine and terrestrial cores.

Investigators finding pollen and spores in smear slides should always

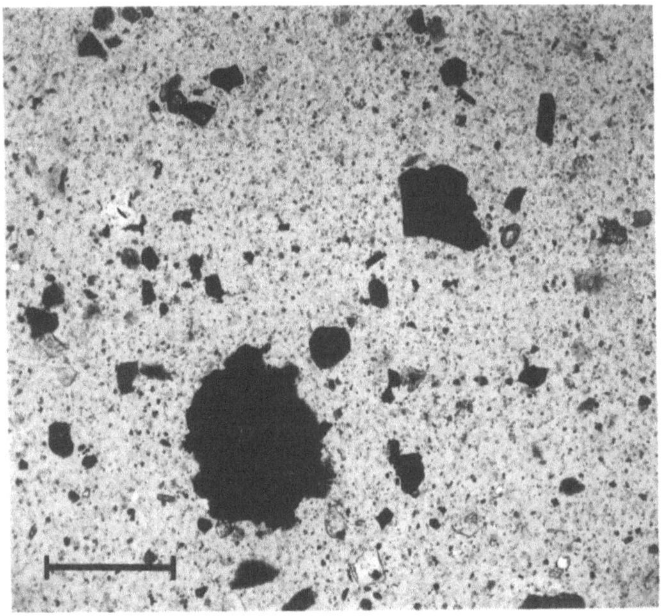

Fig. A5.26. Abundant poorly sorted carbonised organic fragments ('coaly' particles and black debris) and rare pollen and spores in glacial-stage Pleistocene carbonate mud with ice-rafted detritus from the Gardar Ridge, south of Iceland. Plane-polarised light, scale bar equals 0·2 mm. DSDP Leg 94, 611-8-4, 48–50 cm. (Courtesy of Dr Peta Mudie and the Deep Sea Drilling Project.)

remember the possibility of contamination. Spores and pollen from the air may settle on core surfaces during splitting, storage and examination. Further, the cold, damp conditions under which most cores are now stored provided optimal conditions for fungi growth. It is not unusual to see stored cores with large patches of fungi growing on their cut surfaces.

References

BOLLI, H. M., SAUNDERS, J. B. & PERCH-NIELSEN, K. (eds) (1985). *Plankton Stratigraphy*. Cambridge University Press, Cambridge.

BRASIER, M. D. (1980). *Microfossils*. George Allen and Unwin, London.

CLARKSON, E. N. K. (1979). *Invertebrate Palaeontology and Evolution*. George Allen and Unwin, London.

FAIRBRIDGE, R. W. & JABLONSKI, D. (eds) (1979). *The Encyclopedia of Paleontology*. Dowden, Hutchinson and Ross, Stroudsburg, Pennsylvania.

GLAESSNER, M. F. (1972). *Principles of Micropalaeontology*. Hafner

Publishing Company, New York (unchanged reprint of 1945 edition with new Introduction by the author).

HAQ, B. U. & BOERSMA, A. (1978). *Introduction to Marine Micropalaeontology*. Elsevier, New York.

KENNETT, J. (1982). *Marine Geology*. Prentice-Hall, Englewood Cliffs, NJ.

RIEDEL, W. R. (1963). The preserved record: paleontology of pelagic sediments. In *The Sea, Vol. 3,* ed. M. N. Hill. Wiley Interscience, New York, pp. 866–87.

INDEX

Key: Entries in **bold** indicate a major entry on the subject; entries in *italics* refer to an illustration on the page indicated.

MIX
Papier aus verantwortungsvollen Quellen
Paper from responsible sources
FSC® C105338

If you have any concerns about our products,
you can contact us on
ProductSafety@springernature.com

In case Publisher is established outside the EU,
the EU authorized representative is:
Springer Nature Customer Service Center GmbH
Europaplatz 3, 69115 Heidelberg, Germany

Printed by Libri Plureos GmbH
in Hamburg, Germany